Practical Problems in
Groundwater Hydrology

Practical Problems in Groundwater Hydrology

E. Scott Bair
Ohio State University

Terry D. Lahm
Capital University

PEARSON
Prentice
Hall

Upper Saddle River, NJ 07458

Library of Congress Cataloging-in-Publication Data

Bair, E. Scott.
    Practical problems in groundwater hydrology / E. Scott Bair, Terry D. Lahm. .
        p. cm.
    ISBN 0-13-145667-9 (pbk.)
    1. Hydrology--Problems, exercises, etc. 2. Groundwater--Problems, exercises,
etc. 3. Hydrologic models. I. Lahm, Terry D. II. Title.
    GB658.3.B35 2006
    551.49076--dc22

                                                            2005034614

Executive Editor: *Patrick Lynch*
Senior Media Editor: *Chris Rapp*
Project Manager: *Dorothy Marrero*
Editor-in-Chief, Science: *Dan Kaveney*
Executive Managing Editor: *Kathleen Schiaparelli*
Assistant Managing Editor*: Beth Sweeten*
Production Editor: *Shari Toron*
Managing Editor, Science Media: *Nicole M. Jackson*
Media Production Editor: *John Cassar*
Manufacturing Manager: *Alexis Heydt-Long*
Manufacturing Buyer: *Alan Fischer*
Senior Managing Editor AV Management and Production: *Patricia Burns*
Managing Editor, AV Management: *Abigail Bass*
AV Production Editor: *Eric Day*
Art Studio: *LaserWords*
Art Director: *Jayne Conte*
Cover Designer: *Kiwi Design*

© 2006 by Pearson Education, Inc.
Pearson Prentice Hall
Pearson Education, Inc.
Upper Saddle River, New Jersey 07458

Pearson Prentice Hall™ is a trademark of Pearson Education, Inc.

Microsoft®, Excel®, and Windows® are either registered trademarks or trademarks of Microsoft
Corporation in the United States and/or other countries. *Practical Problems in Groundwater
Hydrology* is an independent publication and is not affiliated with, nor has it been authorized,
sponsored, or otherwise approved by Microsoft Corporation.

Printed in the United States of America
10 9 8 7 6 5 4 3 2 1

ISBN 0-13-145667-9

Pearson Education LTD., *London*
Pearson Education Australia PTY., Limited, *Sydney*
Pearson Education Singapore, Pte. Ltd.
Pearson Education North Asia Ltd., *Hong Kong*
Pearson Education Canada, Ltd., *Toronto*
Pearson Educación de Mexico, S.A. de C.V.
Pearson Education—Japan, *Tokyo*
Pearson Education Malaysia, Pte. Ltd.

To our parents,
Dean and Eleanor
and
Dale and Catherine,
whom instilled in their children the
desire to learn.

About the Authors

E. Scott Bair

Scott Bair is Professor and former Chair of the Department of Geological Sciences at The Ohio State University. He received his bachelor's degree in geology from the College of Wooster and his master's and Ph.D. degrees specializing in hydrogeology from the Pennsylvania State University. Following graduate school, Scott worked six years in the Geotechnical Division of Stone & Webster Engineering Corporation before taking a faculty position at Ohio State in 1985. He teaches courses in Water Resources, Hydrogeology, Hydrologic Field Methods, and Computer Modeling of Groundwater Flow and Contaminant Transport. In 1991, he received Ohio State's highest teaching award. He has advised 34 masters and doctoral students. Together, they have worked on a variety of research projects funded by the National Science Foundation, U.S. DOE, U.S. EPA, U.S. Geological Survey, U.S. Department of Agriculture, and a number of state agencies. Scott is a Fellow of the Geological Society of America and during 2000 served as the 23rd Birdsall-Dreiss Distinguished Lecturer sponsored by its Hydrogeology Division. Scott routinely contributes to the teaching of two short-courses for the National Ground Water Association: *Principles of Ground Water Flow, Transport, and Remediation* and *The Design and Analysis of Aquifer Tests.* He has served three times as an Associate Editor of *Ground Water.*

Terry D. Lahm

Terry Lahm is an associate professor of Geology and Environmental Science at Capital University and Director of the Environmental Science Program and Associate Director of the Center for Computational Studies. He received his bachelor's degree in Geology from the College of Wooster and M.S. and Ph.D degrees, specializing in Hydrogeology, from Ohio State University. Between undergraduate and graduate school, Terry worked as a consulting geologist with Engineering-Science, a division of the Parsons Corporation. He joined Capital University's faculty in 1997 where he teaches courses in Hydrogeology, Geomorphology, Computational Environmental Sciences, and Environmental Geology. He has advised numerous undergraduate students in research projects concerning groundwater-surface water interaction, environmental watershed modeling, and water quality issues. Terry is an active member in several scientific professional organizations including the Geological Society of America and the Council on Undergraduate Research and has worked on various projects funded by the National Science Foundation and the W.M. Keck Foundation.

CONTENTS

PREFACE

We created *Practical Problems in Groundwater Hydrology* to reinforce and enhance basic conceptual and quantitative skills of students and professionals through solving real-world problems, analyzing actual field data, and using current image and graphic technologies. Our case study approach to more than 20 exercises from across the United States not only guides users through learning fundamental concepts in groundwater hydrology, but also promotes development of the higher-order thinking skills needed to interpret their analyses and synthesize their knowledge. To engage users' interests, each exercise begins with an intriguing story line extensively illustrated with photographs and diagrams and presented within a multimedia environment on the CD.

Computational templates created in *Excel* worksheets on the CD form the working space for each exercise and the intrinsic mathematical, statistical, and graphical tools in *Excel* are used to solve the necessary equations and create automated graphs and tables. The hydrogeologic concepts, background and technical information, and equations needed to complete each exercise are presented in corresponding chapters in the *Reference Book*, which also includes an *Excel* tutorial and highlighted sections containing *Excel* tips. Several exercises sequentially build on specific case studies so users develop in-depth knowledge of these localities. Embedded macros are used to perform higher-level functions such as particle tracking and complex computations, enabling users to focus on general concepts and interpretation of their results. Each exercise contains questions that guide the user to explore, analyze, and interpret the real-world datasets provided in the exercises. Our emphasis is on understanding the answers to site-specific and overarching scientific questions concerning the central issue in each exercise, not just on making the mathematical calculations.

To create these exercises, we draw on our more than 25 years of collective experiences in teaching groundwater hydrology and hydrogeology courses at both a large research university and a small, primarily undergraduate institution. Many of the story lines, problems, and datasets are taken from our experiences performing consulting projects, working for the U.S. Geological Survey, serving as expert witnesses, teaching short courses for state agencies and professional organizations, and performing hydrogeologic investigations while previously employed by Stone & Webster Engineering Corporation and Parsons Engineering-Science Corporation. To these applied experiences, we add the pedagogic approaches that have been successful in our own classrooms including project-based learning, data exploration, inquiry-based learning, and visualization.

Practical Problems in Groundwater Hydrology is designed for undergraduates and entry-level graduate students taking courses in groundwater hydrology, hydrogeology, water resources, and other related topics. It can be used as a reference book and/or training manual for working professionals. The user is expected to have basic algebra and trigonometry skills, and familiarity with the general concepts introduced in college-level calculus and physical geology. The seven chapters contained herein are a subset of the 12 chapters we originally set out to produce. In the near future, we hope to augment the existing chapters with additional exercises from more varied geologic settings and to add new chapters with exercises on Hydrologic Budgets, Darcy's Law, Heads & Gradients, Springs & Caves, Flow in Fractured Rocks, and Models & Modeling. We believe *Practical Problems in Groundwater Hydrology* provides an exciting and insightful learning environment for the teaching and learning of groundwater hydrology.

Acknowledgments

Attaining our goal of creating a textbook for teaching groundwater hydrology that promotes development of quantitative skills using a framework of applied problems and assembling the myriad materials melded into *Practical Problems in Groundwater Hydrology* was a truly enjoyable endeavor that we could not have accomplished without the encouragement and assistance of friends and colleagues. In these lines we gratefully acknowledge the people who helped us along the way. Our classroom students made many helpful suggestions on the content and framework of the *Excel* worksheets. We are indebted to them for their unwitting willingness to be Alpha testers. Abe Springer, Northern Arizona University; Maura Metheny, McLane Environmental, LLP; Tod Frolking, Denison University; Greg Wiles, College of Wooster, and Donald Pair, University of Dayton gave us encouragement and comments on the initial versions of the *Excel* worksheets and the associated exercises. We appreciate the generosity of Bill Cunningham, U.S. Geological Survey, for providing images and insight, and Marinko Karanovic, S.S. Papadopulos & Associates, for giving us his particle-tracking macro to adapt to our purposes. We are especially grateful for the thoughtful comments from the formal reviewers and Beta testers. Their suggestions improved our materials for use by both faculty and students. Patrick Lynch and Christopher Rapp at Prentice-Hall / Pearson Education guided us through the publication process. We are grateful for their encouragement, expertise, and willingness to publish a book with a new format and learning environment. Ryan Tillett and Sean Welder at Future-Farm Interactive Studio did a magnificent job designing and developing the multimedia environment on the CD and linking it to our *Excel* worksheets. Their creativity and technical knowledge turned our ideas into an exciting visual learning environment. We are indebted to our families for their encouragement and support, which above all, enabled us to take the project to fruition. Last, but certainly not least, we appreciate the fresh air and exercise that our dogs demanded we regularly take and for their uncanny knack at helping us resolve the natty issues that arise during a project.

ESB and TDL

INTRODUCTION

Practical Problems in Groundwater Hydrology weaves two electronic media with this *Reference Book*. The electronic media, which are on the accompanying CD, consist of an interactive multimedia presentation that links to a series of *Excel* worksheets. The interactive presentation enables users to select from 21 exercises in seven chapters that wrap the application of concepts and equations presented in the *Reference Book* around interesting, real-world story lines that are extensively illustrated with images, diagrams, and maps (Figure 1).

Figure 1. Opening screen of interactive CD shows the first level of navigation on the left side of the screen and introductory images on the right side along with the 'EXIT' and full screen toggle buttons in the upper, right corner.

The *Excel* worksheets contain the problem description, datasets, computational templates, and questions for each exercise. The intrinsic mathematical and statistical functions of *Excel* enable users to work through the exercises within the framework of the templates, and the graphical functions of *Excel* are programmed to automatically display computed values for analysis and interpretation. The text writing functions in *Excel* enable users to record their responses to the conceptual questions contained in each exercise. Thus, it is possible to complete the worksheets in a single file. The *Reference Book* presents the relevant concepts, equations, and assumptions needed to complete the exercises and respond to the questions.

We recommend users begin each exercise by examining the background material on the interactive CD and then proceed to the *Excel* file to acquaint themselves with the *Introduction* and *Questions* worksheets. After reviewing this material, the user should read the corresponding chapter and problem in the *Reference Book* to introduce the concepts and equations needed to program the computational templates in the *Excel* worksheets and interpret the results. This degree of entwinement of electronic media with printed media is a new concept; one that we believe fosters quantitative problem-solving skills through exploration, discovery, and analysis of actual datasets from real problems that are extensively described and illustrated.

Below are more detailed descriptions of the *Reference Book*, the interactive CD, and the *Excel* worksheets.

Reference Book

The seven chapters in the *Reference Book* are organized in an identical manner as the interactive CD. The *Reference Book* provides the concepts, principles, and equations needed to perform and interpret the exercises in the *Excel* worksheets. The symbols, units, and terminology used in the equations and descriptions in the *Reference Book* are consistent with those used in the *Excel* worksheets. At the end of each section in the *Reference Book* are (1) a description of the associated exercise, (2) useful *Excel Tips* to facilitate completion of the worksheets, and (3) a table of parameter symbols and names used in the exercise. It is best to consult this information before beginning any exercise. Examples of particularly relevant *Excel Tips* are given below.

Excel Tips

- Periodically save your work to a hard drive to insure it is not lost, swallowed, or sent to blue oblivion.
- Some worksheets use macros to perform mathematical computations and graphical functions. Be sure to set your Macro Security to "Medium." Otherwise, these features will not function.

Interactive CD

The interactive CD is used to introduce users to the background story line and the scientific information needed to complete each of the applied problems. The interactive CD is designed to provide a supportive and interactive environment for visual learning through extensive use of annotated photographs, diagrams, and maps. These materials are to be used in concert with the *Reference Book* and the associated *Excel* worksheets. The interactive CD starts automatically when placed in the computer CD drive and it can be used on both PCs and Macs. It also may be manually started by navigating to the CD drive and double-clicking on the GWHydrology.exe file (see Readme file for more information). The program will determine if *Excel* is present on your computer and will produce a warning if *Excel* is not available and the worksheets cannot be accessed.

The opening screen of the interactive CD (Figure 1) provides introductory images of hydrologic equipment, students and faculty, and field and laboratory activities. The opening screen facilitates navigation to the Table of Contents and also directly to each of the seven chapters. The interactive CD enables users to navigate quickly between the CD and the *Excel* worksheets. A "LAUNCH EXCEL WORKSHEET" button is available at the bottom of every page in each of the chapters (Figure 2).

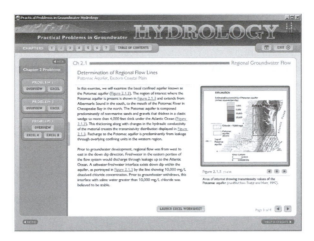

Figure 2.   Example of the interactive CD material for Chapter 2, Problem 1 showing the "LAUNCH EXCEL WORKSHEET" button at the bottom of the screen.

Before launching into an *Excel* worksheet, users are encouraged to read the materials on the interactive CD and view the images, diagrams, and maps that provide instructive background information for each exercise. Instructors may treat the background material as a pre-laboratory exercise for students to complete before arriving in the laboratory to work on the *Excel* worksheets. The *Excel* worksheets also can be launched from the Table of Contents page, which provides a direct link to each of the problems (Figure 3).

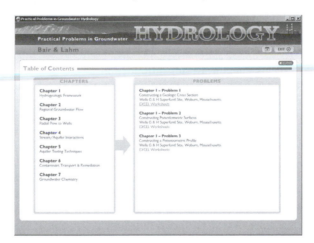

Figure 3.   The Table of Contents on the interactive CD contains the titles of each chapter and exercise, and direct links to the *Excel* worksheets for each problem.

All the figures within the interactive CD can be enlarged by clicking on the button with the plus sign below the figure or double-clicking on the figure itself (Figure 2). Users can advance through the figures independent of the associated text using the forward and backward buttons or click on the hyperlinked text to display the corresponding figure. These mechanisms give users the flexibility to navigate the interactive CD in different ways. The CD window can be switched between full screen and an individual window using the activation button at the upper, right corner of the screen.

Macros are used in five *Excel* worksheets to perform higher-level mathematical computations and graphical functions. This enables users to focus on analysis and interpretation of results. Because the macros are necessary to complete the exercises, it may be necessary for some users to reset the Macro Security setting in *Excel*. To change the macro security level, use the *Tools >> Macro >> Security* command within *Excel* (Figure 4) to set the macro security to "Medium" or "Low." This allows the macros in Chapter 2, Problems 1, 2, 3A, 3B and in Chapter 5, Problem 4 to execute. Once the security level is set to "Medium" or "Low," users can enable the macros in these exercises, which provides the full functionality of the worksheet. Modification of the macro security level must occur before launching the individual worksheets.

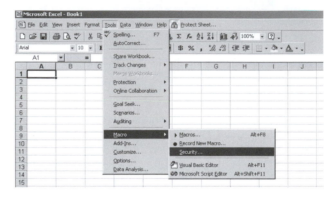

Figure 4. Changing the macro security level in *Excel* is done using the "Tools" menu. The macro security level must be set to "Medium" or "Low" for full functionality of some worksheets.

Excel Worksheets

The *Excel* worksheets contain a brief introduction to the problem, instructions, questions to be answered, and computational templates for performing the quantitative aspects of each exercise. Graphs of pertinent variables are automatically created for interpretation based on the user entering values and programming equations into designated cells in the computational templates. To facilitate ease of use, all of the *Excel* worksheets are designed in a similar layout with the *Introduction* worksheet appearing as the first (leftmost) tab along the menu bar at the bottom of the *Excel* window. The *Introduction* worksheet contains general information about the exercise, relevant images, and instructions on how to complete specific problems (Figure 5). The *Questions* worksheet is accessed from the tabs shown on the bottom menu bar and contains the set of conceptual questions. Responses to the questions can be written within the text boxes embedded in the worksheet or in whatever other form the user or instructor chooses.

All *Excel* worksheets can be printed to facilitate completion of the exercises or evaluation by the instructor. The layout of each worksheet has been designed to maximize the efficiency of printing and includes header information about the chapter and problem identification. The user can adjust the printing parameters for any worksheet, if necessary, by using the Page Setup dialog box under the file menu as seen in Figure 6. Adjustments of the margin settings may be necessary for some printers to avoid awkward pagination issues.

Figure 5. Example of an *Introduction* worksheet to an *Excel* problem showing the tabs to access the other, associated worksheets along the bottom menu bar.

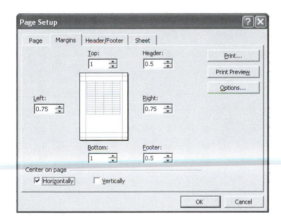

Figure 6. Example of Page Setup dialog box used to change margins for printing *Excel* files.

The associated worksheets contain computational templates and graphs used to complete the exercises. Portions of all the worksheets are protected to prevent accidental modification of pre-programmed cells and graphs. All the computational templates use the same color codes to assist users in determining which cells need to be programmed with intrinsic mathematical or statistical functions, which cells are designated for the user to enter values of specified variables, and which cells are pre-programmed and cannot be altered. These color codes are listed below in Table 1.

Text Color	*Functionality*
Red	Variables input by the user
Blue	Functions to be programmed by the user
Pink	Functions to be programmed by the user
Black	Pre-programmed functions provided to the user

Table 1. Color codes used to define functionality of cells in computational templates.

CHAPTER 1
HYDROGEOLOGIC FRAMEWORK

Principles & Concepts

The problems in this chapter incorporate fundamental geologic and hydrologic concepts concerning characterization of groundwater flow through earth materials and determination of groundwater flow directions and velocities. Construction of site-specific geologic maps and site-specific potentiometric maps provides the information needed to answer the three common questions asked in many groundwater investigations:

- *What are the directions of groundwater flow?*
- *How fast does the groundwater move?*
- *How far will groundwater travel in "x" years?*

Lithologic information from drillers' logs, soil borings, and geophysical logs and surveys provides the basis for constructing plan-view geologic maps, geologic cross sections, isopach maps, and structure contour maps. These maps are used to discern the extent, continuity, thickness, and orientation of the various types of rock and/or sediment that occur at a specific field site. Once the three-dimensional framework of geologic materials is mapped, it is possible to describe the geometry, continuity, connectivity of units, and other characteristics of the aquifers and confining layers that compose the groundwater flow system. Potentiometric surface maps and potentiometric profiles then are constructed and used to characterize the nature of the flow system. Using these graphical tools that depict the geologic and hydrogeologic framework of a specific site, scientists and engineers can address the three basic questions listed above.

Geologic Framework

Geologic maps and cross sections are important tools used to decipher the local geologic and hydrogeologic framework. Creation of these maps commonly occurs in the initial stages of a hydrogeologic study to help provide a conceptual model of the groundwater flow system. A geologic cross section, as seen in Figure 1.1, is an interpretation of the subsurface geologic conditions based on the information available. Because there is never perfect or complete knowledge of the subsurface, some uncertainty is always associated with geologic maps and cross sections. This uncertainty commonly manifests when the mapper extends discontinuous lenses of material beyond and between boreholes, estimates depths to bedrock from regional maps or geophysical surveys, and correlates similar materials between boreholes that do not occur at the same depth. Making realistic geologic cross sections requires experience, practice, and knowledge of how the sediments and bedrock were originally deposited and subsequently modified. Because of this inherent uncertainty, there cannot be an exact depiction of the subsurface conditions at any site. Instead, there are a number of possible, yet similar, depictions that are constrained by the geologic information, the regional geologic setting, and the ways in which the geologic materials were deposited and subsequently modified.

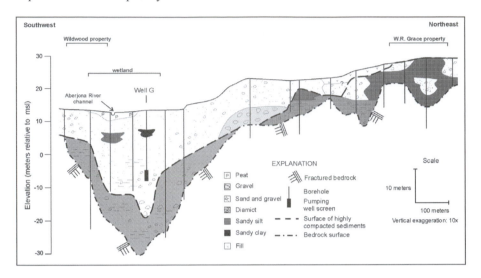

Figure 1.1. Geologic cross section transverse to a buried valley axis showing glacial deposits and bedrock surface (from Metheny, 1998).

Figure 1.2. The hollow-stem auger drilling method pictured here uses steel augers that are "screwed" through the unconsolidated material by the drill rig.

Figure 1.3. Sediment samples are collected using a split-spoon sampling device passed through the hollow center of the steel augers such that sediment is retained inside the split spoon.

The construction of geologic cross sections visually integrates the geologic information obtained from well logs, soil borings, geophysical surveys and logs, outcrops, and shallow pits and trenches. Much of the geologic information used in the following problems is derived from hollow-stem auger drilling (Figure 1.2). In this drilling method, steel augers penetrate the unconsolidated material. Sediment samples are collected during drilling from the drill cuttings, seen at the base of the auger in Figure 1.2, or by using a sampling device passed through the hollow center of the steel augers and pounded into the undisturbed sediment below the lead auger. A split-spoon sampler is commonly used in silts, sands, and gravels. Once extracted, the hollow tube is split longitudinally allowing access to the cored sample (Figure 1.3). A Shelby tube, which is used to sample clays, is a thin cylindrical steel tube that is pushed into the sediment. The sample must be extruded manually in the field or in a laboratory. These types of samples enable geologists to describe changes in lithology with depth and then to correlate zones of similar lithology from well to well consistent with the mode of deposition of the materials. In bedrock, coring is a common method used to collect rock samples for lithologic description. In either rock or sediment, geophysical logs also can be used to delineate vertical lithologic variations in a borehole.

Hydrogeologic Framework

Commonly, the boreholes drilled for lithologic information, as described above, are converted into monitoring wells or pumping wells by the installation of solid casing pipe and well screen into the borehole (Figures 1.4 and 1.5). Techniques used to drill the boreholes are diverse and depend on the objectives of the study, the type of sediment or rock underlying the site, whether the borehole will be converted to a well, and the project budget.

Figure 1.4. Wire-wrapped, 30-slot, stainless steel screen 4-inches in diameter with sample of aquifer material taken with a split-spoon device. Nickel located on material is for scale.

Figure 1.5. The well screen allows water to enter the well only at the depth where it is positioned.

The basic hydrogeologic framework for a given locality also is based on the spatial variation of hydraulic head values measured in the aquifers of interest. This hydraulic head or water-level information is visually integrated in potentiometric surface maps and potentiometric profiles. Potentiometric surfaces show aerial variations in hydraulic head within a single aquifer or layer, whereas potentiometric profiles show spatial variations in hydraulic head along a geologic cross section (Figure 1.6). Potentiometric profiles also allow for the representation of hydraulic heads in multiple aquifers and confining layers. Construction of potentiometric maps and potentiometric profiles provides the fundamental information needed to determine groundwater flow directions, flow velocities, and travel times. Without site-specific water-level data, it is difficult to make any quantitative assessments about a flow system.

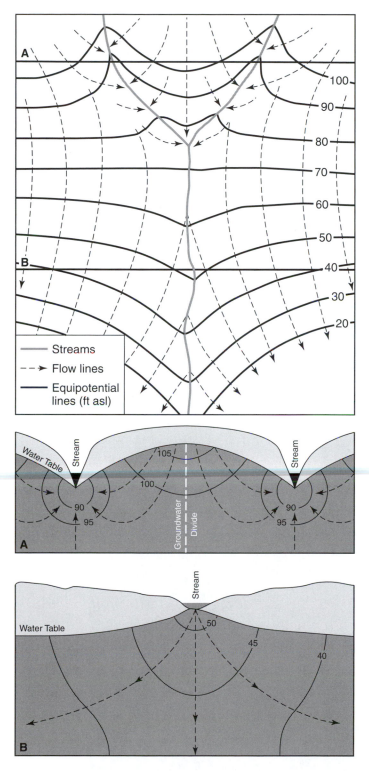

Figure 1.6. Potentiometric surface showing lines of equal hydraulic head (equipotentials) and dashed flow lines with arrows indicating the direction of flow. At cross section A, the streams are gaining water from the aquifer, whereas at cross section B the stream is losing water to the aquifer.

Water-level data are invariably irregularly spaced. Many contouring computer programs re-grid the data and contour the gridded values. These powerful visualization programs produce beautiful maps. They can, however, misrepresent many hydrologic phenomena, particularly if used to contour water levels in an unconfined aquifer that is hydraulically connected to a river, lake, lagoon, or spring. Although a contouring algorithm can be programmed with a spatially and temporally consistent set of biases, it may not properly interpret gaining or losing reaches of streams, inflow and outflow from lakes, or groundwater mounds beneath lagoons unless sufficient well control is available. Hand-drawn potentiometric surfaces of unconfined aquifers commonly are more realistic representations of actual field conditions because hydrologic concepts and local knowledge can be incorporated into the map.

Water levels used in the construction of a potentiometric map should be measured at approximately the same time (synoptic). If water levels from different times of the year or from different years are contoured together, the resulting potentiometric map may contain unrealistic hydraulic gradients produced by the inclusion of seasonal variations in water levels. The fallacious hydraulic gradients may lead to the calculation of incorrect flow velocities and flow directions.

Figure 1.7. Nested wells close to one another completed to two different depths within the same aquifer.

All wells used to construct a potentiometric surface should be screened in the same aquifer and to approximately the same relative depth. If water levels from different aquifers are contoured together, the resulting map may misrepresent directions and magnitudes of hydraulic gradients by including vertical hydraulic gradients from overlying and underlying geologic units. Figure 1.7 shows two wells installed proximal to one another. The well in the foreground is screened in a deep portion of a limestone aquifer, whereas the well in the background is screened in a shallow portion of the same aquifer. Water levels in the two wells are usually different by three or more feet indicating a vertical hydraulic gradient within the aquifer.

To detect these vertical hydraulic gradients within an aquifer, two or more wells close to one another need to be screened at different depths in the aquifer. Wells with short screens (known as piezometers) are preferable when the object is to detect vertical hydraulic gradients. Short screens are more useful because measured water levels are averaged over the length of a well screen. Therefore, the shorter the screened interval, the more precisely we know the vertical location of the water-level measurement. If water levels differ with depth, then vertical gradients exist within the aquifer. The significance of these gradients needs to be evaluated with respect to the project goals. For example, is it worth the added expense to install several wells to evaluate the extent of vertical gradients for a water supply well? The inclusion of vertical gradients in a potentiometric surface map also can lead to errors in the interpretation of flow directions and flow velocities. If water levels do not differ with depth, then vertical gradients do not exist and water levels measured at any depth in the aquifer can be used in the contouring process.

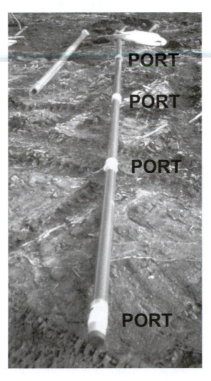

Figure 1.8. Single multiport well casing with four ports, each five feet apart, used to measure water levels and take water samples at different depths within an aquifer.

Figure 1.9. Multiport well casing exposed at the surface after individual plastic tubes were connected to each port visible in Figure 1.8. This system enables water sampling and water-level measurements at discrete vertical locations within the aquifer.

The type of well installed is usually a compromise between the use of a separate borehole for each well or the use of multiport wells in a single borehole (Figures 1.8 and 1.9). There are advantages and disadvantages to each type of installation. Accurate water-level measurements can be made in either type. It is difficult to do realistic hydraulic testing in multiport wells, but water-quality sampling is simplified. Water-quality sampling is more laborious in wells completed in separate boreholes, but any hydraulic testing (as discussed in Chapter 5) will encompass a much larger volume of the aquifer and greater pumping rates can be used. Budget constraints and project goals usually dictate what type of installation is used.

Groundwater hydrologists commonly create contour maps of hydraulic head values to produce potentiometric surfaces (plan view) and potentiometric profiles (cross-section view). The contours on the maps are used to measure hydraulic gradients (change in hydraulic head over change in distance) and to infer groundwater flow directions. The problems in this chapter illustrate how values of hydraulic gradient are combined with values of hydraulic conductivity and effective porosity to estimate groundwater flow velocities and travel times. By solving these problems, you will answer all of the basic questions posed at the beginning of this section.

Problem 1

CONSTRUCTING A GEOLOGIC CROSS SECTION

Wells G & H Superfund Site, Woburn, Massachusetts

Overview

Geologic maps and cross sections are important tools that help us understand the geology and hydrogeology of a field site. These maps allow us to visualize the three-dimensional nature of most groundwater flow systems, which is key to developing a conceptual model of groundwater flow and movement of pollutants. Geologic cross sections are commonly constructed from lithologic descriptions of the earth materials penetrated by soil borings and wells drilled at the site of interest. For example, sediment or rock samples are commonly described by a geologist during the borehole drilling process. Once the borehole is constructed, geophysical instruments can be lowered into the hole and used to measure physical properties of the rock or sediment and pore water. Results of surface geophysical surveys such as radar, resistivity, and seismic can be used to decipher subsurface geologic features.

This exercise introduces the user to the construction of geologic cross sections, which is a fundamental tool in the geoscience profession. Information from boreholes is organized into an illustration of the subsurface geologic framework. This illustration is oriented along a two-dimensional transect through the area of interest; the boreholes represent the locations of known lithologic material. The geologist must then interpolate between these known data points to create a continuous interpretation of the geologic framework. (See the Principles & Concepts section for more background on the geologic framework.)

General Rules and Guidance for Making Geologic Cross Sections

1. *Determine the purpose for the proposed cross section. Is it for a regional understanding of the geology? Is it for determination of a site-specific investigation? The purpose will provide some guidance as to the degree of detail needed.*
2. *Decide on appropriate vertical and horizontal scales that will show the detail desired.*
3. *Locate the well or borehole positions, land surface elevation, well depth, and other known parameters based on field data.*
4. *Transfer the known geologic information from each well log or observation point to the cross section. This information represents the discrete points of knowledge about the subsurface geology. Some important interpretation is*

completed at this step to determine whether geologic information is lumped together under similar descriptors or split into finer distinctions for portraying on the cross section. Remember that the decision to lump or split geologic information is based on the purpose of the cross section and the reliability of the field data used.

5. *Correlate the geologic information between boreholes by incorporating knowledge of the depositional environment of the geologic material. Look for differences in lithology, texture, or sediment or rock properties as a guide to defining contacts between contiguous geologic units. Use solid lines to indicate reasonably certain relations between discrete data points. Dashed lines are used to indicate uncertainty in the nature and location of the contact between distinct geologic units.*

6. *Include a legend to explain the types of geologic materials present.*
7. *Include appropriate orientation and landmark information.*
8. *Include vertical and horizontal scales along with a statement of the amount of vertical exaggeration.*

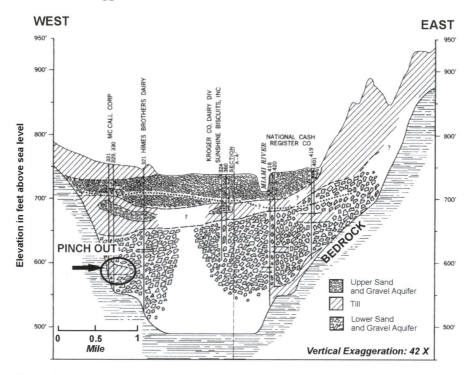

Figure 1.10. West to east cross section of the unconsolidated material underlying the Miami River valley in Dayton, Ohio, showing an interpretation of the geometry and composition of the geologic materials (modified from Norris and Spieker, 1966).

Figure 1.10 shows an example geologic cross section across the buried valley underlying Dayton, Ohio (Norris and Spieker, 1966). Note that the land surface, bedrock surface, and lithologic information are represented in the cross section. Some geologic units are discontinuous (i.e., pinch out) between wells. Additional reading on how this geologic information is collected is available in the Principles & Concepts section of Chapter 1. Site-specific information concerning the geology at the Wells G & H Superfund Site in Woburn, Massachusetts, is available in the Overview of Problem 1 on the CD.

Excel Tips

- In some cases, it is easier to view the introduction worksheets by printing them, for example to see the entire figure presented. You can also decrease the size of the worksheet on your computer screen in order to view more of the worksheet by choosing 75% magnification under the *View >> Zoom* menu.

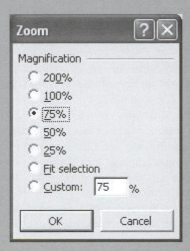

- All formulas programmed into *Excel* start with an equals sign (=) and use standard arithmetic functions denoted by + - * /. For example, "=E16-E17" will subtract the value in cell E17 from the value in cell E16 and place the remainder in the programmed cell.
- Cell anchor: $column$row anchors the cell (column, row) in a formula. Performing a relative copy will maintain the anchored cell within the formula and change only the non-anchored cells (e.g., =$D$12+D2 will anchor cell D12 and not cell D2 during a relative copy).
- Other intrinsic *Excel* functions are listed and explained by clicking the f_x button.
- The cross section template on the "*Cross Section*" worksheet is designed to be printed before creating the geologic cross section by hand. To insure appropriate dimensions of maps, be sure not to highlight the cross section template before printing. If you highlight the maps, *Excel* will print only the chart area and not the legend. Landscape layout is best for printing the geologic cross section in Problem 1.
- ***Do not*** modify the worksheet labeled "*Orientation.*" This map is intended simply to orient the cross section with respect to landmark features. To insure accurate orientation, the coordinate information within the "*Cross Section*" worksheet should not be modified.
- Some worksheets have been protected from unnecessary modification.

Problem 2

CONSTRUCTING POTENTIOMETRIC SURFACES

Wells G & H Superfund Site, Woburn, Massachusetts

Overview

This problem uses the construction of contour lines to portray the three-dimensional potentiometric surface at the Wells G & H Superfund Site in Woburn, Massachusetts. The water levels used to characterize the potentiometric surface were measured in a network of shallow monitoring wells near the two municipal wells, which were closed in May 1979 because of contamination (for additional background information, see the Overview of Problem 1 on the CD). To assess what type of contaminant remediation system would best suit the hydrogeologic conditions at the Superfund site, the U.S. Environmental Protection Agency (EPA) contracted the U.S. Geological Survey (USGS) to perform a long-term aquifer test using wells G and H to determine the degree of interaction between the Aberjona River and the municipal wells.

The first set of measurements taken by USGS scientists was on the morning of December 4, 1985, before former municipal wells G and H began discharging at their historic average rates. The second set of measurements was taken on January 3, 1986, after the former municipal wells had been pumped continuously for 30 days. Both sets of data are from shallow wells whose screened interval penetrates only a short distance (5 to 20 feet) into the aquifer. These data allowed the scientists at the USGS to determine the historic configuration of the water-table surface when wells G and H periodically pumped together between 1967 and 1979. It also allowed scientists to determine historic groundwater flow directions, interactions between the Aberjona River and the underlying aquifer, and rates of groundwater movement. The expert witnesses in the famous federal trial described in the award-winning book *A Civil Action* (Harr, 1995) were able to use these data to help formulate their professional opinions.

People familiar with topographic maps develop an appreciation for the shape of the land surface based on the distribution of contour lines of elevation. Similarly, construction of contour lines is a standard technique used by groundwater hydrologists to visualize the three-dimensional shape of a water table or potentiometric surface. Groundwater hydrologists describe these contour lines as equipotential lines because the contours represent lines of equal potential energy. Listed below are general rules for contouring most types of irregular spatial data, such as land surface elevations, atmospheric pressures, rainfall, snow accumulation, and groundwater levels.

General Rules for Contouring Spatial Data

1. *Use a soft lead pencil and get a good eraser. It will take several tries before you draw a contour surface that honors all the measured data within the mapped region.*
2. *Contour lines connect points of equal value and are usually separated by a constant difference in value, known as the contour interval.*
3. *Contour lines always close on themselves, either within the map or beyond its margins; thus, contour lines do not end or stop in the middle of a map.*
4. *Hachure marks are sometimes drawn on the downhill side of a closed contour line to denote a closed depression.*
5. *Contours are widely spaced on gentle slopes. Contours are closely spaced on steep slopes. Evenly spaced contours indicate a uniform slope.*
6. *A reversal in the direction of slope (such as a valley or ridge) is represented by repeated contours of the same value.*

Constructing Potentiometric Surfaces

Potentiometric surface maps are a special type of contour map. The measured groundwater levels used to construct a potentiometric surface need to be carefully considered. Unlike topographic maps of the land surface, which do not change shape in short periods of time, potentiometric surfaces can change shape in response to pumping stress, seasonal variations in recharge, recharge events, droughts, and other causes. As a result, the timing of the groundwater level measurements is important. All groundwater levels used to construct a potentiometric surface should be measured at the same time or within a short period of time. If groundwater levels from different times of the year or different years are contoured together or groundwater levels from several sets of measurements are averaged, the resulting contours will not represent any actual configuration of the potentiometric surface and likely will lead to misinterpretation of flow directions, hydraulic gradients, flow velocities, and travel times. To eliminate this possibility, it is best to construct a contour map using a synoptic (measured at the same time) set of water-level measurements. So doing insures that transient changes in the potentiometric surface are not inadvertently incorporated into the map.

To represent the "average" or typical position of the potentiometric surface within the normal sinusoidal fluctuation of annual groundwater levels, it is best to make the synoptic set of groundwater level measurements near the midpoint of either the rising limb or the falling limb of the annual hydrograph (Figure 1.11). At these times, groundwater levels should be approximately halfway between seasonal high and low levels. The times of year that the midpoints occur differ across the world depending on climatic factors. In the Midwest of the United States, these typical or "average" times occur in July and January (Sminchak, 1998).

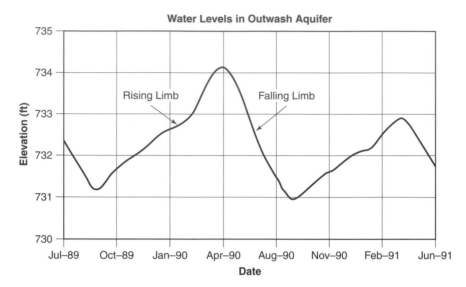

Figure 1.11. Common sinusoidal fluctuation in annual groundwater levels in the Midwest of the United States.

All wells used to construct a potentiometric surface need to be completed in the same aquifer and, if the aquifer is thick, the wells need to be completed to a similar depth in the aquifer. If groundwater levels from different aquifers or from confining layers are contoured together, the resulting contours may misrepresent the actual potentiometric surface because unintended vertical gradients may be included in the map. To detect vertical gradients, one or more pairs of laterally adjacent wells need to be completed at different depths within the aquifer. If heads vary with depth, then vertical gradients are present and their significance needs to be evaluated. If heads are nearly the same at different depths, then the depth of the well is not important and wells completed at any depth in a single aquifer can be used in construction of the potentiometric surface. It is particularly important to eliminate vertical hydraulic gradients from the construction of a water-table map. Because there is normally a downward component of flow at the water table, it is necessary to limit wells used in the construction of a water-table map to those wells with screens that straddle the water table and penetrate only a short distance into the aquifer. If wells with deeper screens are used, the resulting contour map may include unintended vertical gradients.

The contour interval used to portray the potentiometric surface is commonly a constant value, but it need not be. It can be useful to use a smaller contour interval in areas with flatter gradients and a larger contour interval in areas with steeper gradients. If changes in the contour interval are made within a single map, it is important to include this information in the legend on the map. Equipotential lines should not extend beyond the geographic limits of the well network without being designated as inferred. This delineation is usually done by drawing the contours as dashed lines. Listed below are several rules for constructing potentiometric surface maps.

Guidance for Making Potentiometric Surface Maps

1. *All wells used to represent a single potentiometric surface need to be completed in the same aquifer to approximately the same depth.*
2. *Use synoptic sets of groundwater level and surface water data.*
3. *Equipotential lines do not cross or intersect.*
4. *Where there is hydraulic connection between a river or stream and an underlying unconfined aquifer, equipotential lines may make a horseshoe bend upstream or upvalley across gaining streams and a similar bend downstream or downvalley across losing streams.*
5. *Equipotential lines representing a groundwater mound form closed contours.*
6. *Equipotential lines around a pumping well form a closed depression, known as a cone of depression, which is usually **not** denoted with hachure marks.*
7. *Cones of depression in a sloping potentiometric surface are asymmetrical; closer spaced contours lie on the downgradient side and wider spaced contours on the upgradient side.*

Many software programs contour irregular spatial data and can be used to produce beautiful maps. However, these programs may misrepresent many hydrologic phenomena, particularly if used to contour groundwater levels from an unconfined aquifer that is hydraulically connected to lagoons, lakes, rivers, or springs. Although contouring algorithms can be programmed with spatially and temporally consistent bias, unlike drawing contours by hand, contouring algorithms may not properly interpret gaining or losing reaches of streams, inflow and outflow from lakes, groundwater mounds under lagoons, or spring catchments unless sufficient well control is present. As a result, hand-drawn water-table surfaces of unconfined aquifers can be more representative of actual field conditions because hydrologic concepts and local knowledge can be incorporated into a hand-drawn map.

Drawing Flow Lines on a Potentiometric Surface

Flow lines trace the path that hypothetical particles of water take through a groundwater flow system under steady-state conditions. In an isotropic aquifer, where the hydraulic conductivity is the same in every direction, flow lines intersect equipotential lines at right angles. This right-angle rule recognizes that the steepest hydraulic gradient between two adjacent equipotential lines is along the perpendicular distance between them. Thus, flow lines drawn on a potentiometric surface are depicted by tracing lines that are continuously perpendicular to the set of equipotential lines. Flow-line construction assumes steady-state conditions, as with a single set of synoptic water levels. If recorded groundwater levels vary with time, then the paths that water particles take through the flow system also change with time. In most cases, it is assumed that the potentiometric surface represents the "average" or typical configuration of the flow system. In so doing, it is recognized that some fluctuations in groundwater levels and flow directions occur and some deviations in flow lines occur in response to these fluctuations. As a result, these fluctuations may cause the flow directions, hydraulic gradients, and flow velocities to vary.

Flow lines are an important concept in delineating the movement of groundwater within the saturated zone, particularly for flow systems in which there is little or no vertical movement of groundwater and flow is predominantly two-dimensional and horizontal. It is important to recognize that a contoured potentiometric surface is a two-dimensional representation of a three-dimensional surface. As such, flow lines drawn on a potentiometric surface always appear horizontal and cannot depict any vertical movement (even if it is present) of groundwater (Figure 1.12). Vertical components of groundwater flow are best depicted by constructing a potentiometric profile (see Problem 3 in this chapter).

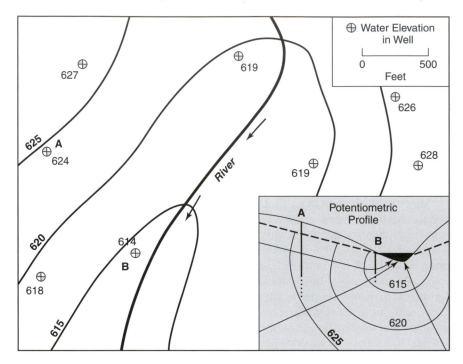

Figure 1.12. Equipotential lines allow for the depiction of the water table or potentiometric surface and are expressed in plan view and in profile view (shaded inset). Only flow lines drawn on a potentiometric profile portray the vertical movement of groundwater.

Measuring Hydraulic Gradients

Hydraulic gradients are vectors and thus possess both magnitude and direction. The magnitude and direction of the hydraulic gradient are estimated, regionally, by using flow lines and equipotential lines on a potentiometric surface and are calculated, locally, by using groundwater levels measured in at least three wells (see Problem 3 in this chapter). The magnitude of the hydraulic gradient indicates how much potential energy is available to move water at a point in the flow system. The hydraulic gradient *(i)* is calculated using the following equation.

$$i = \frac{Difference \;\; in \;\; Hydraulic \;\; Head}{Flow \;\; Line \;\; Length} = \frac{dh}{L} \qquad (1\text{-}1)$$

On a regional basis, *dh* is measured on a potentiometric surface map as the contour interval between two adjacent equipotential lines. *L* is measured on the same map as the scaled distance between the two equipotential lines. On

Figure 1.12, the magnitude of the hydraulic gradient between points A and B is illustrated in the following equation.

$$i = \frac{10\,feet}{1000\,feet} = 0.01\,feet\Big/feet = 0.01 \qquad (1\text{-}2)$$

Note that this hydraulic gradient would be the same if the difference in hydraulic head and the flow line length were given in meters instead of feet (the hydraulic gradient is unitless). For many hydraulic gradients, the magnitude is less than 1.0. On Figure 1.12, the direction of the hydraulic gradient is along a line segment perpendicular to the 620- and 625-foot equipotential lines. As demonstrated in this figure, however, the distance L measured on the potentiometric surface map does not include the vertical component of the flow line shown on the potentiometric profile. As a result, the measurements taken from the potentiometric surface slightly underestimate L (the length of the flow line) and this causes the magnitude of the hydraulic gradient (i) to be slightly overestimated. Investigations can be performed to insure that the three-dimensional nature of flow lines is accounted for in any analysis of flow gradients where a high degree of accuracy is warranted (Abriola and Pinder, 1982).

Groundwater Flow Velocities

The velocity of groundwater flow is directly related to the hydraulic gradient. The magnitude of the velocity vector is calculated using a modified form of Darcy's Law, and the direction of the velocity vector in most instances is the same as that of the hydraulic gradient vector. The velocity of groundwater flow commonly is called the average linear flow velocity because it represents the average or bulk velocity of groundwater, not the velocity of individual particles of water that may travel faster or slower than the average velocity. The equation for computing the average linear flow velocity is given by

$$V_L = \frac{K\,\dfrac{dh}{L}}{n_e} \qquad (1\text{-}3)$$

where

V_L is the average linear flow velocity (L/T),
K is the hydraulic conductivity (L/T),
dh is the difference in hydraulic head (L),
L is the flow line length between points of hydraulic head measurement (L), and
n_e is the effective porosity (expressed as a decimal).

The quantity dh/L is the hydraulic gradient and is either computed from a potentiometric surface or is computed using the three-point procedure described in Problem 3 in this chapter. Site-specific values of K can be computed using a variety of techniques, many of which are described in Chapter 5. An example of this calculation, which is based on the data from Figure 1.12 and assumes a hydraulic conductivity of 100 feet/day and a porosity of 15%, is shown below.

$$V_L = \frac{(100\,feet\,/\,day)(0.01\,feet\,/\,feet)}{(0.15)} = 6.7\,feet\,/\,day \qquad (1\text{-}4)$$

When this equation is used, it is important to make all distance and time units consistent (e.g., if hydraulic heads and flow line lengths are in feet, the hydraulic conductivity must be in terms of feet per unit time).

Once the groundwater velocity is determined, the travel time can be calculated between two points in the groundwater flow system using the following equation.

$$Travel\,time = \frac{Length\,of\,Flow\,Path}{Average\,Linear\,Flow\,Velocity} \qquad (1\text{-}5)$$

For example, in the area represented by Figure 1.12 the distance between points A and B is 1000 feet and the average linear flow velocity is 6.7 feet/day. Therefore, the travel time between these two points in the groundwater flow system is 1000 feet / 6.7 feet/day or about 150 days.

Excel Tips

- The maps in this problem (Dec 4 '85 P-map and Jan 3 '86 P-map) are designed to be printed before contouring is performed by hand. To insure appropriate dimensions of maps, do not highlight the maps before printing. If you highlight the maps, *Excel* will print only the chart area and not the legend. The map may also have an incorrect orientation. Portrait layout is best for the potentiometric surface in Problem 2.
- There are hidden worksheets labeled "*Coordinates Dec*", "*Coordinates Jan*." Data in these worksheets are used within the automatic plotting routines and do not need to be modified.
- Some worksheets have been protected from unnecessary modification.

Parameter	Definition
dh	Difference in hydraulic head (L)
i	Hydraulic gradient (L/L)
K	Hydraulic conductivity (L/T)
L	Flow line length between points of hydraulic head measurement (L)
n_e	Effective porosity (expressed as a decimal)
V_L	Average linear flow velocity (L/T)

Table 1.1 Parameter Definition Table – Chapter 1, Problem 2

Problem 3

CONSTRUCTING A POTENTIOMETRIC PROFILE

Wells G & H Superfund Site, Woburn, Massachusetts

Overview

This problem uses contour lines to display the three-dimensional shape of the water table in the glacial outwash aquifer in a portion of the Aberjona River valley and adjacent uplands in Woburn, Massachusetts. The groundwater levels were measured in a network of shallow monitoring wells surrounding municipal wells G and H, which were closed in May 1979 because they contained concentrations of organic solvents in excess of U.S. EPA and state standards. As discussed in Problem 2, two sets of synoptic groundwater level measurements were made. The focus of this exercise is the second set of data, which was taken on January 3, 1986, after the two municipal wells had pumped continuously for 30 days at their average historical rates of 700 gallons per minute (gpm) (well G) and 400 gpm (well H). Both wells were built on mounds of artificial fill above a riparian wetland in the Aberjona River floodplain. The wells are each approximately 90 feet deep and 2 feet in diameter, and they have 10-foot-long screens below 80 feet of casing. As such, the wells partially penetrate the outwash sands and gravels in the buried valley aquifer, which is up to 130 feet deep near the wells (see Problem 1).

The potentiometric surface constructed from groundwater level measurements in the shallow wells used in Problem 2 cannot by itself portray the complete three-dimensional hydrologic picture. Potentiometric profiles provide a view of groundwater flow in the vertical plane, which cannot be portrayed in a two-dimensional, plan-view potentiometric surface. This exercise uses some of the same wells used in Problem 2, but it also uses groundwater level measurements from wells screened at intermediate and greater depths in the buried-valley aquifer. Most of these wells were drilled in clusters of two, three, and four wells in a small area, as shown in Figure 1.13, to measure changes in water levels and contaminant concentrations with depth.

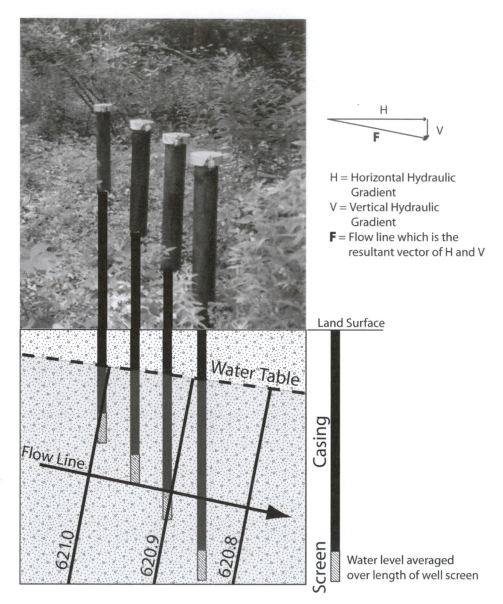

Figure 1.13. Nested wells completed in the glacial aquifer at Woburn, Massachusetts. The shallowest well has a higher water level than the deeper wells. Thus, groundwater flow has a downward component in this area.

Clustered (nested) wells completed with screens at different depths enable examination of changes in hydraulic head with depth. As a result, vertical components of flow can be portrayed on a potentiometric profile. Although the general technique of drawing contour lines is the same on a potentiometric profile as on a potentiometric surface, construction of a potentiometric profile requires additional considerations. Listed on the next page are guidelines for drawing potentiometric profiles.

Guidance for Making Potentiometric Profiles

1. *All equipotential lines originate at the water table and intersect it only at points where the value of the equipotential line is the same as the elevation of the water table (e.g., the 625-foot equipotential line intersects the water table only at points where the water table has an elevation of 625 feet).*
2. *Groundwater mounds, which can occur beneath hills and beneath losing reaches of rivers and streams, are represented by equipotential lines that curve downward and by flow lines that diverge from the groundwater mound (Figure 1.6).*
3. *Groundwater troughs, which can occur beneath gaining reaches of rivers and streams and around partially penetrating pumping wells, are represented by equipotential lines that curve downward and flow lines that converge on the riverbed or partially penetrating well screen (Figure 1.6).*
4. *Potentiometric profiles, like geologic cross sections, are commonly drawn with an exaggerated vertical scale. Although this graphical technique enables more vertical detail to be shown, in most cases it invalidates the right-angle rule of intersections between flow lines and equipotential lines.*
5. *Equipotential lines and flow lines that cut across geologic materials with substantially different hydraulic conductivity are refracted (see Problem 2 in Chapter 2). The degree of refraction is difficult to display precisely because potentiometric profiles commonly are drawn with an exaggerated vertical scale.*

Constructing Potentiometric Profiles

Drawing equipotential lines and flow lines on a potentiometric profile can be complex and thought provoking. This is particularly true when partially penetrating rivers and pumping wells are present, as at the wells G and H site. The 10-foot-long well screens that penetrate about two-thirds of the total depth of a buried valley will cause equipotential lines to form "halos" around the well screen (see the Overview of Problem 3 on the CD). These equipotential lines will terminate in pairs at the water table on opposite sides of the cone of depression, which in cross section looks like an asymmetrical funnel centered about the pumping wells (see Figure 1.3.5 in the Overview of Chapter 1, Problem 3 on the CD).

Calculation of hydraulic gradients by using a potentiometric profile is more complex than on a potentiometric surface, because most cross sections are vertically exaggerated. Hydraulic gradients are calculated using the same concepts used in Problem 2, but the equation now has to account for the exaggerated vertical scale.

Determination of flow lines within a potentiometric profile differs slightly from the plan-view interpretation. Because the flow system may be composed of various hydrostratigraphic layers with differing hydraulic conductivities, flow lines will refract at the interface of two such hydrostratigraphic units. The Law of Refraction is used to guide the drawing of equipotential lines and the resulting flow lines (see Chapter 2 in *Reference Book*). A second factor complicating the construction of flow lines is that cross sections, like the one in this exercise, are usually drawn with vertical exaggeration. That exaggeration allows each hydrostratigraphic unit to be drawn at a reasonable scale.   However, the vertical exaggeration nullifies the "right angle rule" described previously. Thus, equipotential lines and flow lines will not necessary cross perpendicular to one another. Although there are graphical techniques that can be used to adjust for vertically exaggerated profiles, they are beyond the scope of most introductory textbooks.

Excel Tips

- Landscape layout is best for the potentiometric profile in Problem 3.
- Some worksheets have been protected from unnecessary modification.

Parameter	Definition
dh	Difference in hydraulic head (L)
i	Hydraulic gradient (L/L)
K	Hydraulic conductivity (L/T)
L	Flow line length between points of hydraulic head measurement (L)
n_e	Effective porosity (expressed as a decimal)
V_L	Average linear flow velocity (L/T)

Table 1.2 Parameter Definition Table – Chapter 1, Problem 3

CHAPTER 2
REGIONAL GROUNDWATER FLOW

Principles & Concepts

This chapter discusses the concepts of regional groundwater flow and examines the basic processes that control the movement of groundwater on scales of tens to hundreds of kilometers through complex geologic terranes. Understanding these concepts and processes and demonstrating their application to water resource sustainability and the isolation of nuclear waste and hazardous liquid waste are the focus of this chapter.

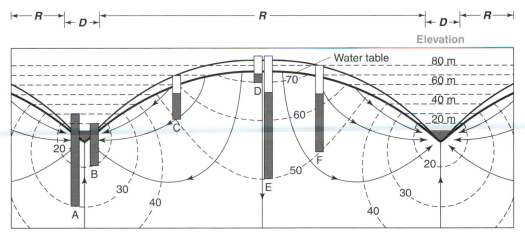

R = Recharge Area
D = Discharge Area

Figure 2.1. Cross section through an idealized unconfined aquifer showing the orientation of equipotential lines (dashed lines) and flow lines (solid lines with arrows). Recharge occurs at topographic highs where the water table is elevated and the hydraulic gradient is downward. Discharge occurs at topographic lows where the water table is low and the hydraulic gradient is upward. (Modified from Hubbert, 1940; used with permission of the University of Chicago Press.)

As seen in previous exercises, groundwater flows from areas of high hydraulic head to areas of low hydraulic head. From a regional perspective, the major driving force behind differences in hydraulic head is the topography of the land surface. The water table generally mimics the land surface topography in a subdued fashion. The result is the development of high elevation recharge areas (locations where water enters the flow system) and low elevation discharge areas (locations where water leaves the flow system). Recharge areas in unconfined aquifers commonly lie on topographic highs where infiltrating precipitation reaches the water table. Vertical hydraulic gradients in recharge areas are downward, forcing water deeper into the flow system. As seen in Figure 2.1, well D is completed to a shallow depth and has a higher hydraulic head than an adjacent well E, which is completed in a deeper portion of the aquifer. Therefore, the equipotential lines and flow lines indicate downward flow in the recharge area. Discharge areas in unconfined aquifers commonly lie in topographic lows where groundwater discharges to streams, seeps, lakes, or springs. Vertical hydraulic gradients are generally upward in discharge areas. In Figure 2.1, well A is completed in the deep portion of the aquifer and has a higher hydraulic head than adjacent well B, which is completed in a shallow portion of the aquifer. As a result, the equipotential lines and flow lines indicate upward flow in the discharge areas. Note that recharge occurs through much of the land surface in Figure 2.1, whereas discharge is focused in smaller areas under surface water features such as a stream.

In confined aquifers, recharge and discharge areas are not as strictly controlled by topography. A confined aquifer can receive recharge by direct infiltration in unconfined portions of the aquifer and also by vertical leakage across adjacent confining layers. Discharge from confined aquifers can occur as seepage to the surface through unconfined portions of the aquifer, well extraction, and/or vertical leakage across adjacent confining layers. Leakage across confining layers is commonly an important control on the overall hydrodynamics of a regional flow system and on changes in water quality (Figure 2.2). In this illustration from Long Island, New York, we see the interactions of precipitation, ocean water, freshwater, and saltwater within the hydrologic cycle and the regional groundwater flow system. The topography, geology, and fluid density of saltwater and freshwater influence the movement of groundwater in this flow system.

Equipotential lines and flow lines, as seen in Chapter 1, help scientists and engineers understand the movement of groundwater at a local scale. Potentiometric surfaces and potentiometric profiles also can be used on a regional scale to interpret regional patterns of groundwater flow. As seen in previous exercises, in isotropic media flow lines intersect equipotential lines at right angles because groundwater moves down the steepest hydraulic gradient from high hydraulic heads to low hydraulic heads. Groundwater flow velocities and travel times of hypothetical water particles can be estimated for a regional system if the average linear flow velocity is estimated using equation 1-3. From this velocity information, the bulk travel time of groundwater is determined with the following relation.

$$Travel\ time = \frac{Distance\ along\ flow\ line}{Average\ linear\ flow\ velocity} \qquad (2\text{-}1)$$

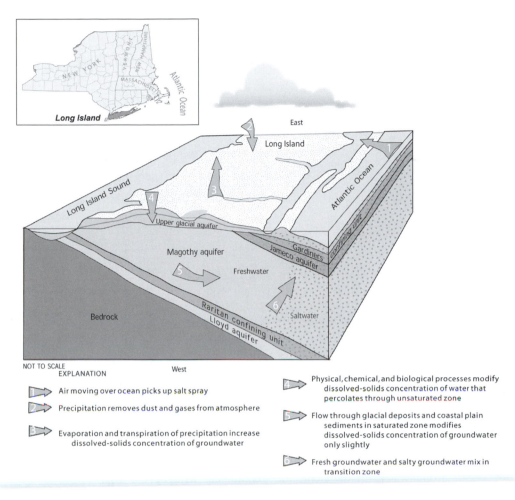

Figure 2.2. Regional hydrologic cycle on Long Island, New York, where infiltration of precipitation recharges the upper glacial aquifer and Magothy aquifer. Cross-formation flow across confining layers allows leakage between the Lloyd, Magothy, and Jameco aquifers and discharge to the Atlantic Ocean (modified from Olcott, 1995).

On a regional scale, three-dimensional groundwater flow may be complex and is controlled, in part, by the nature of the geologic materials and structures. Often, we conceptually simplify the geologic framework and the flow system into a series of aquifers and confining layers through which groundwater percolates from recharge areas to discharge areas. Figure 2.3 illustrates the groundwater flow system along a cross section of Long Island, New York. Water recharges the flow system through the upper glacial aquifer at topographic highs and travels downward and horizontally through the Magothy aquifer. Some of the regional flow is refracted downward across the Raritan confining unit, which consists of silt and clay, and travels into the Lloyd aquifer. Flow discharges from the Lloyd aquifer to Long Island Sound completing the regional flow pattern that is controlled by both topography and the hydrogeologic framework.

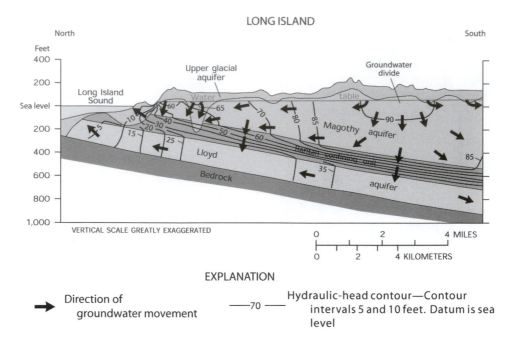

LONG ISLAND

Figure 2.3. Geologic and potentiometric profile along north-south cross section through Long Island, New York, showing geologic units and regional hydraulic-head distribution (modified from Olcott, 1995).

Figure 2.4 shows the potentiometric surface of the Magothy aquifer and illustrates several regional groundwater flow concepts explored further in this chapter. The hydraulic-head distribution indicates that regional recharge occurs along the central portion of Long Island and creates an east-west-trending groundwater divide along the longitudinal axis of the island. Regionally, groundwater flow is radial away from the groundwater highs that correspond with topographic highs (here, they are end moraines) on the island. Groundwater north of the divide flows north to discharge into Long Island Sound, whereas groundwater south of the divide flows south to discharge into the Atlantic Ocean. The effect of groundwater pumping on the regional potentiometric surface can be observed in Queens County, on the far west side of the island, where a closed depression formed in the potentiometric surface. Sustainability of groundwater resources is also a topic addressed in the problems of Chapter 2.

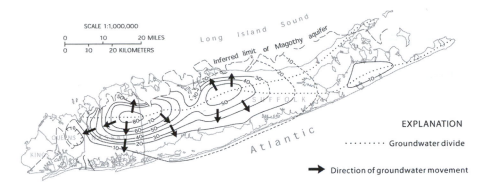

— 20 ---- Potentiometric contour—Shows altitude at which water level would have stood in tightly cased wells during March–April 1984. Dashed where approximately located. Hachures indicate depression. Contour interval 10 feet. Datum is sea level.

Figure 2.4.   The potentiometric surface in the Magothy aquifer underlying Long Island, New York, showing groundwater flow radiating away from regional potentiometric highs underlying glacial end moraines near the center of the island. Note the cone of depression in Queens County, on the far west end of the island, caused by groundwater abstraction for New York City (modified from Olcott, 1995).

Problem 1

DETERMINATION OF REGIONAL FLOW LINES

Potomac Aquifer, Eastern Coastal Plain

Overview

The regional movement of groundwater within the Lower Cretaceous Potomac aquifer in the Coastal Plain of Virginia and North Carolina is controlled by the regional geologic framework and hydraulic conditions. This regional aquifer system has been extensively used during the past century to meet the growing demand for freshwater resources in the area. Extraction of groundwater has had a significant effect on the regional flow field throughout this 10,000-square-mile region on the eastern coast of the United States.

Effect of Pumping on Regional Flow

Pumping water from an aquifer draws down the potentiometric surface and creates a local cone of depression centered on the well. The lowering of water levels propagates outward from the well until an equilibrium is reached between the amount of water withdrawn from the aquifer and the amount of water supplied to the aquifer through recharge or leakage. The area within the aquifer that supplies water to the pumping well is commonly called the capture zone. The shape of the capture zone depends on many factors, which we will explore more fully in subsequent exercises. Examples of the cones of depression and capture zones of pumping wells are seen in Figure 2.5 at West Point and at Franklin. The cones of depression are characterized by equipotential lines with hachures indicating a closed depression in the contoured surface. The flow lines illustrate the general direction of groundwater flow within the regional aquifer. For our purposes, this potentiometric surface represents steady-state conditions in the aquifer when the water levels were measured in 1980. In reality, these water levels vary over time as recharge fluctuates from variations in infiltration and leakage across confining units into aquifers and from changes in discharge through wells (Figure 2.5).

When two or more cones of depression abut, they create a local groundwater divide between them (Figure 2.6). Groundwater on one side of the divide flows down the hydraulic gradient to one well, whereas groundwater on the other side of the divide flows down the hydraulic gradient to the other well. These groundwater divides represent portions of the boundaries delineating the capture zones of the two wells. In regional flow systems, cones of depression and corresponding groundwater divides may extend for several kilometers around pumping wells.

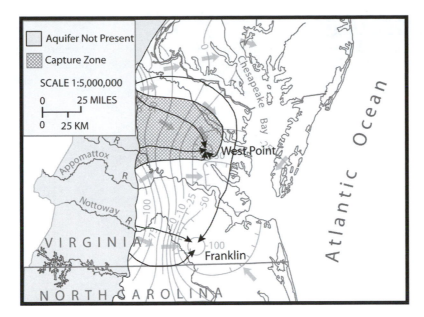

Figure 2.5. Potentiometric surface of the Potomac aquifer with shaded capture zone of West Point pumping center and flow lines (arrows) showing groundwater flow directions to Franklin pumping center. Hachures indicate a closed depression (modified from Trapp and Horn, 1997).

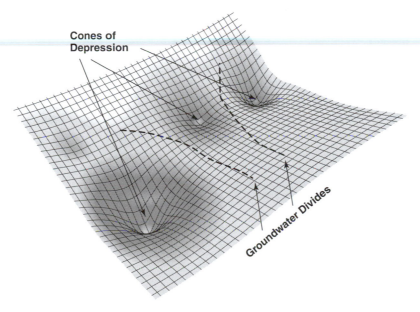

Figure 2.6. Three pumping wells create three individual cones of depression that overlap, thus producing groundwater divides between the cones of depression.

Excel Tips

- This worksheet contains a macro to make calculations. Your security setting within *Excel* should be set to "Medium," which will allow you to run associated macros. The security setting can be modified within *Excel* under the Tools >> Macro >> Security menu, which brings up the dialog box shown below.

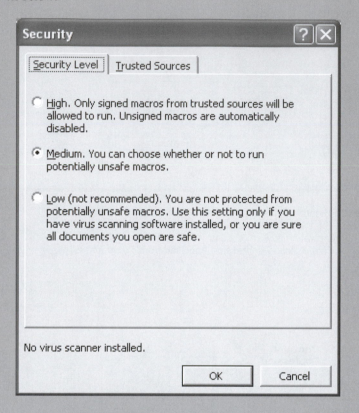

- If the security setting within *Excel* is set to "Medium," the following dialog box will appear when you try to open the worksheet. To allow full functionality of the worksheet, select "Enable Macros" in this dialog box, as seen below.

- The numerical model and particle-tracking algorithm used in this exercise to display flow lines do not need to be modified. Therefore, the *VISUAL*

BASIC code and much of the numerical model are protected and cannot be accessed by the user.

- The groundwater flow model must iterate to the hydraulic head solution. To ensure that the iteration option is activated in *Excel*, use the Tools >> Options menu and click on the Calculation tab. The iteration box should be checked and assigned a maximum number of iterations of 1000 and a maximum change of 0.001, as shown below.

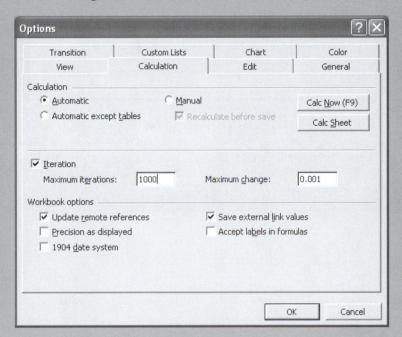

- When choosing the starting location of particles, be sure to use the mouse and click on the active cells shown in light blue on the "*Particle*" worksheet.

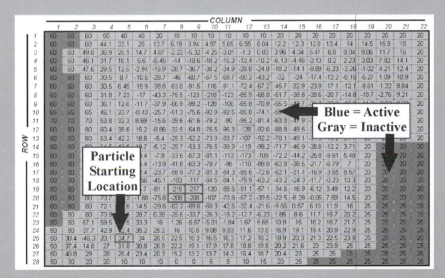

- After choosing the particle's starting location by using the mouse to highlight the cell, you will be asked for the time step to use in the model (see illustration below). We suggest using time-step values from 5,000 to 10,000

days. The larger the time step, the faster the particle tracking will be completed but the less accurate will be the solution. Experiment with various time steps to see which is optimal with your computer system.

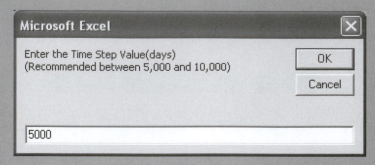

- If the simulation fails to complete, use the "Esc" key to stop the simulation and choose the "End" button to stop execution of the *VISUAL BASIC* code.
- Remember to save your work often. *Excel* has an AutoSave function available under the Tools menu. The AutoSave function may be an Add-In function with some versions of *Excel*.

Problem 2

REFRACTION OF GROUNDWATER FLOW AT GEOLOGIC INTERFACES

Black Mesa Basin, Northeast Arizona

Overview

Many hydrogeologic studies show that the flow lines and equipotential lines refract as groundwater flows across boundaries between geologic materials of different hydraulic conductivity (Hubbert, 1940). Figure 2.7B illustrates this change in equipotential lines and the subsequent change in flow lines in layered rock units. As seen in the figure, flow lines refract or bend at the geologic interface causing water to flow more vertically through the lower permeability material of the confining layer. The refraction of groundwater flow has substantial ramifications on the travel time of water through regional flow systems and ultimately on the final discharge locations of the groundwater.

Scientists and engineers recognize that groundwater flow patterns are modified at geologic interfaces where different hydraulic conductivities exist in adjacent layers or strata. The amount of refraction of flow lines and equipotential lines at these interfaces is proportional to the hydraulic conductivity ratio of the two layers. This relation is analogous to that described by Snell's Law for the refraction of light at density interfaces. In groundwater flow the trigonometric relations dictate that the ratio of the tangent of the incidence angle to the tangent of the refraction angle is equal to the hydraulic conductivity ratio of the two layers (Figure 2.8). This relation is expressed in the following equation.

$$\frac{\tan \theta_1}{\tan \theta_2} = \frac{K_1}{K_2} \qquad (2\text{-}2)$$

where

θ_1	is the angle of incidence in Layer 1 (degrees),
θ_2	is the angle of refraction in Layer 2 (degrees),
K_1	is the hydraulic conductivity of Layer 1 (L/T), and
K_2	is the hydraulic conductivity of Layer 2 (L/T).

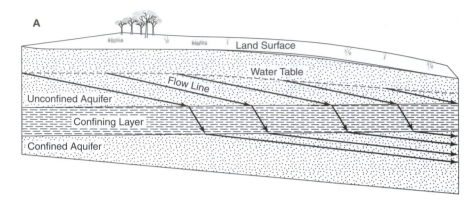

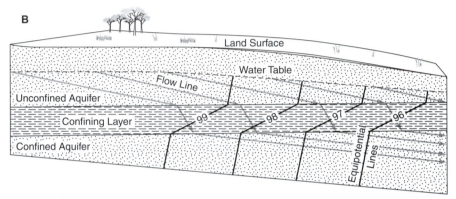

Figure 2.7. (A) Flow lines (arrows) traversing an unconfined aquifer, a confining layer, and a confined aquifer. (B) Refraction of flow lines (arrows) and equipotential lines (solid lines) through a low-permeability layer (modified from Heath, 1998).

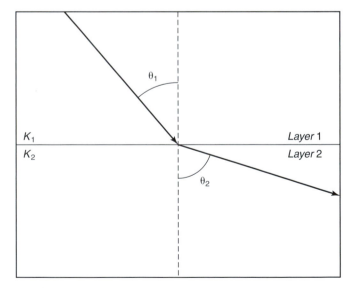

Figure 2.8. Refraction of a groundwater flow line at the interface between two geologic materials with different hydraulic conductivities (K_1 and K_2). Dashed line represents a normal to the geologic interface.

Given the angle of incidence and the hydraulic conductivity ratio of the two layers, we can solve the above equation for the angle of refraction.

$$\theta_2 = \arctan\left(\frac{K_2}{K_1}\tan\theta_1\right) \tag{2-3}$$

The equipotential lines also refract at the interface between Layer 1 and Layer 2 such that the flow line segments in each layer remain perpendicular to the equipotential line segments in each layer.

Implications for Regional Groundwater Flow

In this exercise we explore simple patterns of groundwater flow, refraction of groundwater flow, and the sensitivity of flow patterns to changes in hydraulic conductivity. Flowing groundwater encounters variations in lithology during the movement from recharge areas to discharge areas within a groundwater flow system. The interface between each change in rock and/or sediment type is a potential location where flow lines and equipotential lines can refract. In large sedimentary basins composed of alternating aquifers and confining layers, widespread refraction can substantially affect regional directions of groundwater flow, travel times, locations of recharge and discharge areas, and contaminant pathways.

Excel Tips

- This worksheet contains a macro to make calculations. Your security setting within *Excel* should be set to "Medium," which will allow you to run associated macros. The security setting can be modified within *Excel* under the Tools >> Macro >> Security menu, which brings up the dialog box shown below.

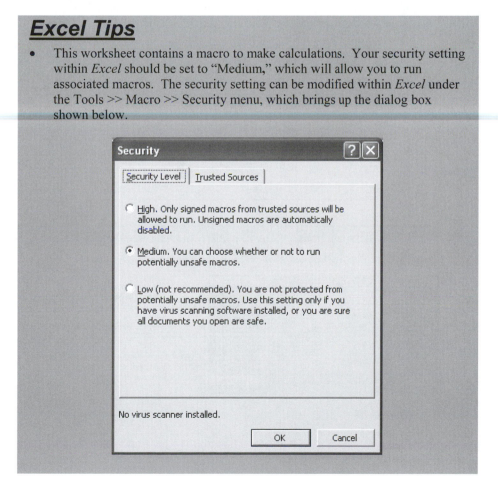

- If the security setting within *Excel* is set to "Medium," the following dialog box will appear when you try to open the worksheet. To allow full functionality of the worksheet, select "Enable Macros" in this dialog box, as seen below.

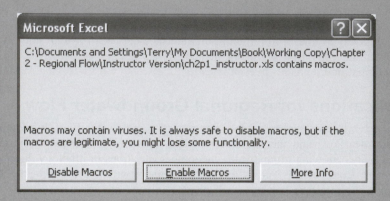

- Radians() converts degrees to radians.
- Tan() returns the tangent in radians.
- Atan() returns the arctangent in radians.
- Degrees() converts radians to degrees.
- Remember to save your work often. *Excel* has an AutoSave function available under the Tools menu. The AutoSave function may be an Add-In function with some versions of *Excel*.
- Enter values of hydraulic conductivity (*K*) for each layer in units of feet per day.

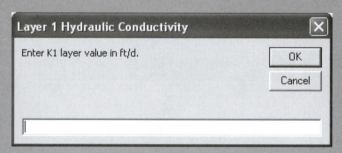

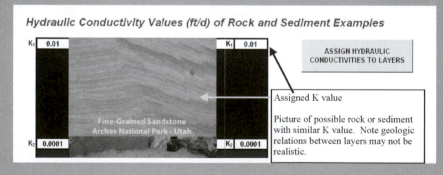

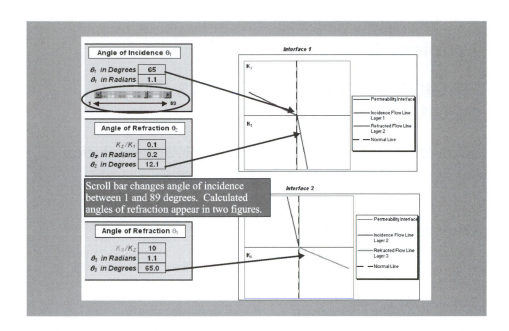

Parameter	Definition
K_1	Hydraulic Conductivity of Layer 1 (L/T)
K_2	Hydraulic Conductivity of Layer 2 (L/T)
K_3	Hydraulic Conductivity of Layer 3 (L/T)
θ_1	Angle of Incidence in Layer 1 (degrees)
θ_2	Angle of Refraction in Layer 2 (degrees)
θ_2	Angle of Incidence in Layer 2 (degrees)
θ_3	Angle of Refraction in Layer 3 (degrees)

Table 2.1 Parameter Definition Table – Chapter 2, Problem 2

Problem 3

REGIONAL FLOW PATTERNS IN SEDIMENTARY BASINS

Hazardous Waste and Radioactive Waste Isolation

Overview

A common way to dispose of hazardous liquid wastes is to inject them into deep saline geologic units. In the United States, this practice is regulated by the U.S. EPA Underground Injection Control Program. The goal of the program is to ensure isolation of the millions of tons of hazardous liquid wastes that are pumped into deep subsurface formations each year by the more than 500 Class I injection wells permitted to operate in the United States (see the problem Overview on the interactive CD). Isolation of radioactive wastes from nuclear power plants is also a significant environmental management challenge. These radioactive wastes remain toxic to humans and the environment for tens of thousands of years. Isolation of this material from the accessible environment is of critical importance to society. Understanding all the facets of the many forms of waste isolation is a complex scientific, engineering, political, and societal problem. Understanding how regional groundwater flow patterns are influenced by the geologic framework helps us better understand some of the major factors that control waste migration in the subsurface.

Linear and Sinusoidal Water-Table Configurations

In Part A of this exercise, we examine regional groundwater flow patterns similar to those described by M.K. Hubbert (1940) (Figure 2.1). Flow systems develop with downward hydraulic gradients in recharge areas and upward hydraulic gradients in discharge areas. József Tóth (1963) showed how to simulate simple regional groundwater flow patterns in large sedimentary basins by computing two-dimensional, cross-sectional equipotential distributions in a flow system with a sinusoidal water-table configuration. Figure 2.9 shows the formation of local, intermediate, and regional flow systems. These benchmark papers markedly advanced the understanding of how regional groundwater flow systems operate under various hydrogeologic conditions. In this exercise, we will recreate some of the flow patterns presented in these papers using an automated numerical flow model and particle-tracking macro similar to that used in Problem 1 in this chapter.

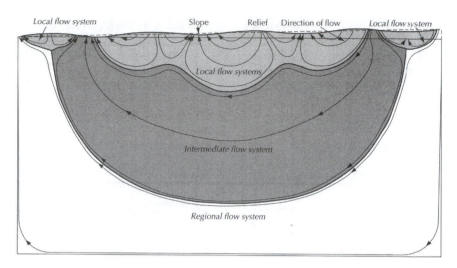

Figure 2.9. Local, intermediate, and regional groundwater flow systems develop where the water table has a sinusoidal shape and a regional hydraulic gradient (Tóth, 1963). Local flow systems are shallow and are characterized by adjacent recharge and discharge areas. Intermediate flow systems have recharge and discharge areas that are not adjacent. Regional flow systems result when groundwater traverses the entire extent of the basin. (Tóth, 1963; Copyright 1963 American Geophysical Union. Reproduced/modified by permission of American Geophysical Union).

The water-table configuration in our simulations of the hydraulic head distribution uses equation 2-4 (Domenico and Schwartz, 1998).

$$h(x, z_0) = \left[z_0 + \frac{B'x}{L} + b \sin\left(\frac{2\pi x}{\lambda} \right) \right] \qquad (2\text{-}4)$$

where

$h(x, z_0)$ is the hydraulic head of the water table at any x coordinate and elevation above the datum z_0 (L),

z_0 is the elevation at the lowest point of the water table (L),

B' is the height above z_0 of the highest point of the water table (L),

x is the x coordinate distance (L),

L is the basin length (L),

b is the amplitude of the sine wave (L), and

λ is the wavelength of the sine wave (L).

In this formulation, the regional hydraulic gradient (i) is described by the following equation.

$$i = \frac{B'}{L} \qquad (2\text{-}5)$$

This equation allows for the representation of both a linear and sinusoidal water table. If the amplitude of the sine wave (b) is set to zero, then the third term of the equation drops out and the water table becomes linear. All the variables used in equation 2-4 for our simulations are defined in the worksheets, which are used to simulate the hydraulic head distributions in a variety of geologic settings. Figure 2.10 shows some of the geometric parameters used in making these automated simulations.

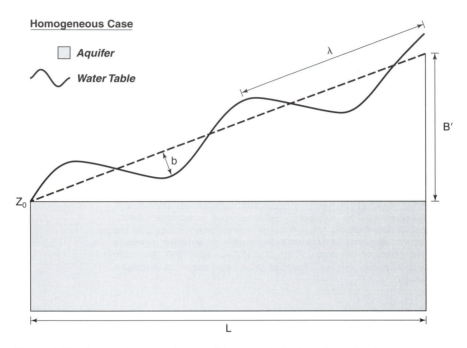

Figure 2.10. Geometric parameters used to compute the two-dimensional flow field for linear and sinusoidal water-table configurations.

Geologic Complexities

In Part A of this exercise, the automated simulations assumed a uniform geologic framework, and thus hydraulic conductivity is assumed to be isotropic and homogeneous. In Part B, we add some geologic complexity to the automated simulations so you can better understand how refraction of flow influences the regional groundwater flow patterns. First, we simulate a layered geologic system where there are three aquifers separated by two confining layers. This heterogeneous framework will change the flow patterns within the regional flow system. Figure 2.11 illustrates a cross section of an idealized, layered regional groundwater flow system. Some of the groundwater discharging to the river travels deep through the regional flow system and may be several centuries to millennia old, whereas other groundwater discharging to the river travels along a shallow flow path and may be only a few years to tens of years old.

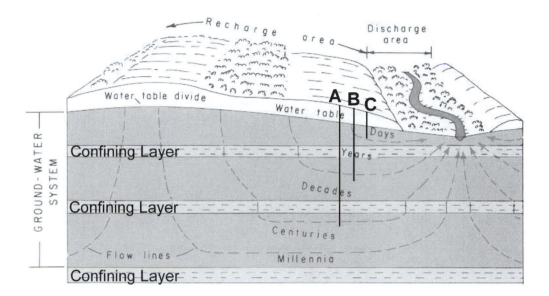

Figure 2.11. Regional groundwater flow lines showing movement from a recharge area to a discharge area in an idealized geologic framework with three aquifers and three confining layers. Well A accesses older water in the deep aquifer, well B accesses water in the intermediate aquifer, and well C accesses younger water in the shallow aquifer (modified from Heath, 1998).

The final geologic complexity represented in our automated simulations is a highly conductive (high permeability) lens within a regional flow system. As is true of a layered system, a lens also has a large effect on regional flow patterns. These simulations will provide insight into the types of regional flow patterns that can develop in different geologic settings and how refraction of groundwater at geologic interfaces can have a large regional effect. These regional flow patterns simulated in Parts A and B show the importance of carefully characterizing the geologic framework of a basin before locating a radioactive waste repository or a hazardous-waste injection well.

Excel Tips

- This worksheet contains a macro to make calculations. Your security setting within *Excel* should be set to "Medium," which will allow you to run the associated macros. The security setting can be modified within *Excel* under the Tools >> Macro >> Security menu, which brings up the dialog box shown below.

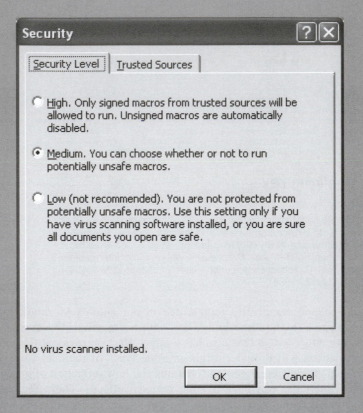

- If the security setting within *Excel* is set to "Medium," the following dialog box will appear when you try to open the worksheet. To allow full functionality of the worksheet, select "Enable Macros" in this dialog box, as seen below.

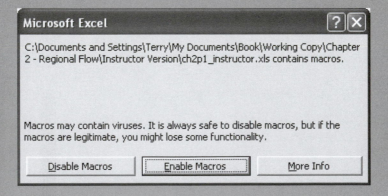

- The numerical model and particle-tracking algorithm used in this exercise to compute hydraulic heads and display flow lines do not need to be modified. The *VISUAL BASIC* particle-tracking code and much of the flow model are protected and cannot be accessed by the user.

- With your mouse, choose one cell within the flow field matrix (light gray cells) before clicking the "Calculation" button. If you do not choose an active cell, you will receive the following error message.

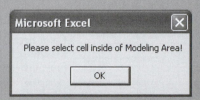

- The flow line calculations can be aborted at any point by hitting the "Esc" button. The following dialog box will appear. Choose the "End" button to abort the simulation.

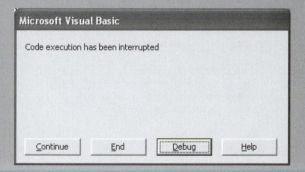

- The regional flow model must iterate to the hydraulic head solution. To ensure that the iteration option is activated in *Excel*, use the Tools >> Options menu and click on the Calculation tab. The iteration box should be checked and assigned a maximum number of iterations of 1000 and a maximum change of 0.001, as shown below.

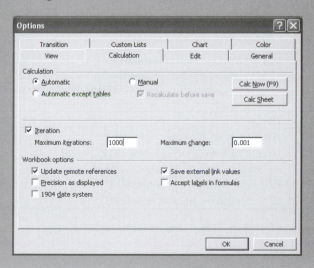

- When choosing the starting location of the particles, be sure to use the mouse to click on the active cells in the flow field matrix, which are shown in light gray on the worksheet and are assigned row and column numbers.

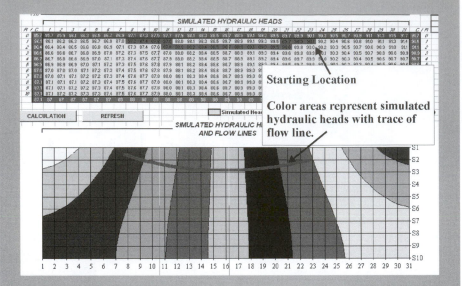

- Depending on the speed of your computer, the calculation of flow lines may take a few moments to a minute or more. The traces of the flow lines are displayed on the hydraulic head distribution as they are calculated. To speed up these calculations, close all nonessential programs including internet browsers.
- Remember to save your work often. *Excel* has an AutoSave function available under the Tools menu. The AutoSave function may be an Add-In function with some versions of *Excel*.
- Once the flow line simulation is complete, a travel time dialog box will pop up displaying the total travel time of the particle in days and years.

Parameter	Definition
b	Amplitude of the sine wave (L)
B'	Height of the major topographic high (L)
$H(x, z_0)$	Hydraulic head of the water table (L)
i	Regional hydraulic gradient (L/L)
λ	Wavelength of the water table (L)

Table 2.2 Parameter Definition Table – Chapter 2, Problem 3

CHAPTER 3

RADIAL FLOW TO WELLS

Principles & Concepts

As seen in previous chapters, pumping wells can dramatically change groundwater flow velocities and flow directions. Measurement of the magnitude of change is usually quantified by the amount and areal extent of drawdown created by a pumping well. Groundwater moves into a pumping well in response to the lowering of the hydraulic head in the well relative to the surrounding aquifer. This lower hydraulic head in the well is caused by the removal of water from the wellbore by a pump, bucket, or some other device. The removal of water from the well creates a hydraulic gradient in all directions that is inward toward the well (radial flow). In this view, a pumping well is a vertical shaft extending into the aquifer to a depth below the water table (unconfined aquifer) and creates a cone of depression in the water table or potentiometric surface around the pumping well (Figure 3.1).

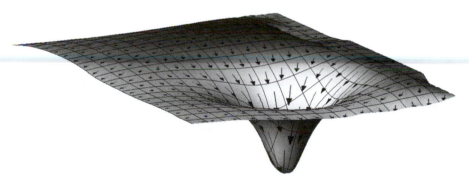

Figure 3.1. Simulated cone of depression showing potentiometric surface around a pumping well and vectors of hydraulic gradient.

A variation of this concept is the special case of a flowing artesian well, whose hydraulic head in the well is naturally higher than the land surface. This relation causes water to flow out of the well without the aid of a pumping device (Figure 3.2). Flowing artesian wells are becoming far more rare as groundwater resources become depleted and potentiometric surfaces decline below land surface elevations.

Figure 3.2. Discharge from a flowing artesian well at Prairie Du Chien, Wisconsin, in 1885 (T.C. Chamberlin, 1885).

Groundwater hydrologists have developed several quantitative methods for predicting the drawdown around a pumping well. These equations commonly are used for regulatory compliance to estimate how future pumping will alter groundwater levels. The equations also are commonly used by expert witnesses in court to estimate what effect pumping may have had on water levels, especially in cases where there are few historic water-level measurements. For example, groundwater hydrologists may want to forecast the effect of a new industrial, irrigation, or municipal well on existing groundwater users before the new well is constructed. Another example is the use of these equations to design a dewatering or depressurizing system that will be put to use before excavation of earth for a large foundation under a building, bridge, or tunnel.

This chapter focuses on some of the quantitative methods used to predict the response of an aquifer to an applied pumping stress. In making these predictions, we simulate the response of the aquifer to the pumping stress using a mathematical model of the groundwater flow system. The model is a representation of the groundwater flow system in mathematical terms. It combines the geology of the site with appropriate equations describing the physics of groundwater flow.

The mathematical groundwater flow models presented in this chapter are in the form of radial flow equations having continuous variables in space and time that describe the aquifer, well characteristics, and pumping stress at the site of interest. These mathematical models require simplifying assumptions concerning the geology and the nature of the pumping stress. This simplification is needed because real conditions are more complex than those represented by the equations. For example, to use the mathematical model the geology commonly must be simplified but still be sufficiently realistic to calculate a reasonable aquifer response to the pumping stress for the intended purpose.

Equations describing groundwater flow to a well can be classified according to whether the equation describes steady-state flow or transient flow to the well. Under steady-state conditions, water levels in the aquifer no longer change with time in response to pumping. This lack of change implies that the areal extent of the cone of depression developed around the pumping well is in equilibrium with the pumping stress and that the amount of water produced by the well is balanced by an equal amount of water entering the aquifer as recharge or as leakage across confining layers. If steady-state conditions exist, hydraulic gradients do not vary with time, although they may vary from place to place within the aquifer. Steady-state conditions also imply that there is no net change in the amount of water in storage, because water levels in the aquifer are neither rising nor falling. Thus, steady-state equations of groundwater flow have no storage coefficient term and no time term.

Under transient conditions, water levels change with time. Thus, hydraulic gradients can change temporally and spatially as water moves into or out of storage. Transient equations like the Theis equation contain storage coefficient and time terms and describe the change in drawdown around a pumping well.

The shape of the cone of depression around a pumping well is influenced by both the aquifer itself and by the construction and operation of the well. If the well screen penetrates the entire saturated thickness of aquifer, then the well is said to be fully penetrating. If the well screen penetrates only a portion of the aquifer, then the well is said to be partially penetrating. In fully penetrating wells, equipotential lines are vertical and flow lines are radially horizontal. Thus, the same head is measured at the top as at the bottom of a vertical line through the aquifer. In a partially penetrating well, equipotential lines curve and flow lines must also curve to maintain their orthogonal relation. As seen in Figure 3.3, hydraulic heads near the partially penetrating pumping well can differ substantially depending on the depth of measurement. Therefore, the Theis equation becomes a poor predictor of drawdowns near a partially penetrating pumping well because some of its assumptions are violated.

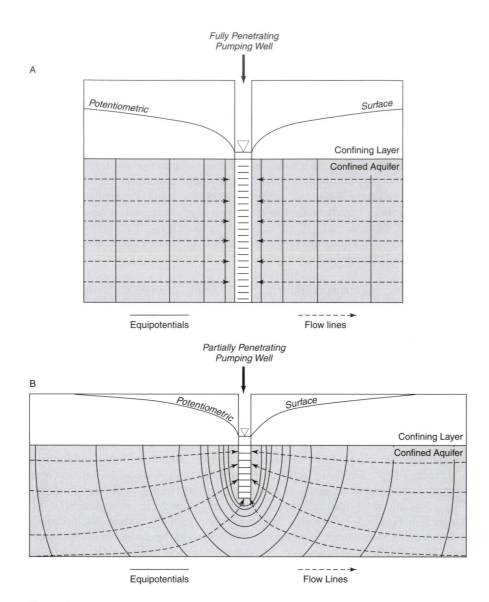

Figure 3.3. (A) A fully penetrating well produces horizontal flow lines and vertical equipotential lines. (B) A partially penetrating well produces vertical equipotential lines and horizontal flow lines only in regions distant from the pumping well. Near the pumping well, the equipotential lines curve in response to the length and position of the partially penetrating well screen. Hydraulic heads will differ with depth where the curvature of the equipotential lines is substantial.

Problem 1

THEIS EQUATION FOR DISTANCE VERSUS DRAWDOWN

Fire Protection Well, River Bend Station Nuclear Power Plant, St. Francisville, Louisiana

Overview

Radial flow of groundwater to a pumping well can be mathematically expressed by combining various forms of Darcy's Law with equations of continuity. Flow to wells in confined or unconfined aquifers and under steady-state or transient conditions can be addressed mathematically. One of the most basic methods and simplest sets of conditions describes a confined aquifer with transient flow to a well that penetrates the entire saturated thickness of the aquifer (i.e., the well is fully penetrating). Theis (1935) was the first to represent these conditions in mathematical terms. Figure 3.4 shows flow to a fully penetrating well in a confined aquifer where the flow is horizontal and radial.

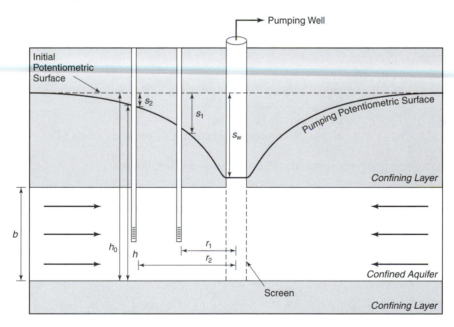

Figure 3.4. Conceptualization of radial flow to a fully penetrating pumping well showing initial and pumping potentiometric surfaces. Two observation wells at radii r_1 and r_2 illustrate the spatial variation in drawdowns s_1 and s_2. The saturated thickness of the aquifer is b and drawdown in the pumping well is s_w.

In 1935 C.V. Theis published his original paper, which described the analogous transient behavior of heat flow to a line sink in an infinite conductive solid and groundwater flow to a well in an infinite aquifer. The mathematical formulation he presented described the amount of drawdown (s) created by a pumping well at any time (t) after pumping commences, at any radial distance (r). This mathematical model became known as the Theis equation and requires certain assumptions about the nature of the aquifer, the pumping rate, and the design of the well. The Theis equation is as follows

$$s = h_o - h = \frac{Q}{4\pi T} \int_u^\infty \frac{e^{-u}}{u} \, du \qquad (3\text{-}1)$$

where

s	is drawdown at any time and distance from the pumping well (L),
h_0	is initial hydraulic head at any distance [$t = 0$] (L),
h	is hydraulic head at the same distance after elapsed time [$t = t$] (L),
Q	is discharge from pumping well (L^3/T),
T	is the transmissivity of the aquifer (L^2/T), and
u	is the Theis equation parameter.

The Theis equation parameter (u) is defined as

$$u = \frac{r^2 S}{4Tt} \qquad (3\text{-}2)$$

where

r	is radial distance from the pumping well to any distance (L),
S	is storage coefficient (L^3/L^3),
T	is transmissivity (L^2/T), and
t	is elapsed time since the beginning of pumping stress (T).

The assumptions inherent in the Theis equation are as follows:

Aquifer - *isotropic, homogeneous, uniform thickness, flat lying, infinite in areal extent, overlain and underlain by impermeable layers, releases water instantaneously from storage.*

Well - *fully penetrates the aquifer, discharges at a constant rate, no borehole storage.*

The Theis equation (equation 3-1) cannot be integrated directly but can be approximated using the following expansion.

$$s = h_0 - h = \frac{Q}{4\pi T}\left[-0.5772 - \ln u + u - \frac{u^2}{2 \cdot 2!} + \frac{u^3}{3 \cdot 3!} - \frac{u^4}{4 \cdot 4!} + \; \ldots \right] \qquad (3\text{-}3)$$

To simplify the expression, the entire series expansion is usually denoted by the term $W(u)$, which is called the Theis well function. Thus, the Theis equation can be written in a simplified form.

$$s \; = \; h_0 - h \; = \; \frac{Q}{4 \pi T} \; W(u) \tag{3-4}$$

where

> $W(u)$ is the Theis well function, and
>
> u is the Theis equation parameter and is the argument of the function.

We use the series expansion (equation 3-3) of the integral to solve the Theis equation with *Excel*. (See Appendix D for a list of $W(u)$ versus u values). This first exercise entails programming the Theis Well Function $W(u)$ into the "*Distance-Drawdown*" worksheet and solving the Theis equation for drawdown (s) at various values of time (t) and distance (r). Because each successive term in the infinite series becomes smaller and smaller, $W(u)$ is approximated by accounting for a finite number of terms. The separate terms are then summed to obtain an approximate value of the integral. The number of terms we use differs between problems but is usually six to ten. We limit the approximation to ten terms in this exercise to illustrate the nature of the Theis well function.

The Theis equation is a powerful tool for estimating the effects of future and previous pumping stress on nearby water levels. These predictions are predicated on the Theis equation being a realistic model of the actual flow system. To determine whether the Theis equation is a realistic mathematical model, the user must check to see if the simplifying assumptions required of the Theis equation compare favorably with the actual conditions at the site. Many of the assumptions apply only to the area influenced by the pumping stress. For example, the aquifer must be isotropic and homogeneous throughout the area influenced by the pumping stress. If all the assumptions are met, then calculations based on the Theis equation may be reasonable. If the assumptions are not met, then use of the Theis equation may yield erroneous estimates.

Another check on the suitability of using the Theis equation to make predictions can be made with the results from a multiple-well aquifer test. This test is a controlled field experiment performed to determine whether observed time-drawdown behavior matches theoretical time-drawdown behavior based on the Theis equation. If the observed response matches the response predicted by the Theis equation, then use of the Theis equation as a mathematical model should produce reasonable results, at least in the short term. The aquifer test data also can be used to estimate site-specific values of transmissivity and storage coefficient in other mathematical models that are used to make predictive calculations. More information about aquifer tests and their interpretation is found in Chapter 5.

An additional aquifer characteristic used to describe the ability of water to move through an aquifer is transmissivity. Transmissivity (T) is the rate of flow under a unit hydraulic gradient through a cross section of the aquifer that has a unit width for the entire saturated thickness of the aquifer. It is determined by the following equation.

$$T = K\,b \qquad\qquad (3\text{-}5)$$

where

T is the transmissivity of an aquifer (L^2/T),
K is the hydraulic conductivity (L/T), and
b is the saturated thickness of the aquifer (L).

It is tempting to use the Theis equation to compute the drawdown in the pumped well itself using a radius representing the size of the wellbore or the size of the sand filter or gravel pack. This approach may greatly underestimate the drawdown that actually occurs in the pumped well because of additional drawdown caused by frictional losses created as water moves through the sand filter or gravel pack and across the screen. The Theis equation does not account for this additional drawdown, which is collectively known as well loss. It is possible to estimate the amount of additional drawdown in the pumped well owing to well losses for a given pumping rate on the basis of analysis of a step-drawdown test performed in the pumping well (Kruseman and de Ridder, 1990).

Excel Tips

- FACT(n) is an intrinsic *Excel* function that takes the factorial of the number "n." Use FACT(n) to increase the efficiency of programming the W(u) function.
- PI() inserts the constant π into an equation.
- LN(n) calculates the natural logarithm of the number "*n*."
- SUM() sums the cells designated within the parentheses.
- Cell anchor: $column$row anchors the cell (column, row) in a formula. Performing a relative copy will maintain the anchored cell within the formula and change only the other non-anchored cells in the formula (e.g., =$D$12+D2 will anchor cell D12 and not cell D2 during a relative copy).
- Exponent symbol: ^ raises a number to the designated power (e.g., 3^2 = 9).
- Other intrinsic *Excel* functions are listed and explained by clicking the f_x button in the Toolbar.
- Using the F2 key on a cell with a formula will highlight the reference cells used in the formula. This allows you to more easily evaluate the accuracy of the formula programmed into the cell.
- *Excel* will give an error message if a programming error occurs such as illustrated below with missing parentheses. However, *Excel* will not necessarily give an error message if the formula is entered incorrectly.

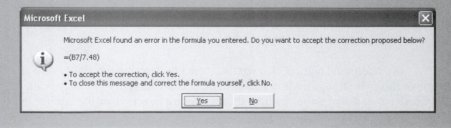

- "#DIV/0!" will appear in cells if a division by zero error has occurred. #DIV/0! displays if the denominator in any formula is zero including reference cells that may also contain an error. Thus, one "#DIV/0!" error may propagate through the worksheet to multiple cells. Check your formulas for accuracy to avoid these errors.
- Save your work often to avoid losing information.

Parameter	Definition
b	Saturated thickness of the aquifer (L)
h_0	Initial hydraulic head at any distance (L)
h	Hydraulic head at the same distance after elapsed time (L)
K	Hydraulic conductivity of the aquifer (L/T)
Q	Discharge from pumping well (L^3/T)
r	Radial distance from the pumping well to any distance (L)
S	Storage coefficient (L^3/L^3)
s	Drawdown at any time and distance from the pumping well (L)
T	Transmissivity of the aquifer (L^2/T)
t	Elapsed time since the beginning of the pumping stress (T)
u	Theis equation parameter
$W(u)$	Theis well function

Table 3.1 Parameter Definition Table – Chapter 3, Problem 1

Problem 2

THEIS EQUATION FOR TIME VERSUS DRAWDOWN

Dewatering System, River Bend Station Nuclear Power Plant, St. Francisville, Louisiana

Overview

The temporary lowering of the water table in shallow aquifers to facilitate excavation of rock or sediment before construction of foundations for buildings, bridges, and construction of tunnels is a common practice. This problem examines the dewatering of an unconfined shallow aquifer in Louisiana for the construction of a nuclear power plant. The River Bend Station power plant was constructed northwest of Baton Rouge, Louisiana, outside the town of St. Francisville. In order to excavate the 41-acre footprint of the foundation for the buildings to a depth of approximately 100 feet, the water table needed to be lowered 65 feet to facilitate excavation and construction in dry conditions. The water table was lowered by installing 44 high-capacity pumping wells around the perimeter of the excavation and discharging the pumped water to a local bayou (see the Overview of Problem 2 on the CD).

The Theis equation can be used to predict the long-term behavior of water levels in both confined and unconfined aquifers if certain assumptions are valid. These assumptions include the Theis assumptions regarding the aquifer and the pumping well that are described in Problem 1 of this chapter and for the long-term behavior of water levels in unconfined aquifers. In unconfined aquifers, the Theis equation can be a reasonable predictor of long-term water levels. The late-time drawdown response to a pumping well in an infinite unconfined aquifer is consistent with the Theis assumptions except that the late-time response is controlled by the specific yield, not by the coefficient of storage (Kruseman and de Ridder, 1990).

Well Interference

Pumping a well causes water levels in the aquifer to draw down and form a conical depression in the water table of an unconfined aquifer or in the potentiometric surface of a confined aquifer.

If two pumping wells are in proximity, their individual cones of depression may overlap. In this case, the calculated drawdowns from each well are additive in areas where the cones overlap, such that the total drawdown at any location in the aquifer is the sum of that produced by the two pumping wells (Figure 3.6). The additive nature of drawdowns is known as well interference and refers to the composite drawdown produced by two or more pumping wells. For example, if well A created 3 feet of drawdown halfway between wells A and B, and well B created 2 feet of drawdown at this same point, then the composite (total) drawdown would be 5 feet at this halfway point.

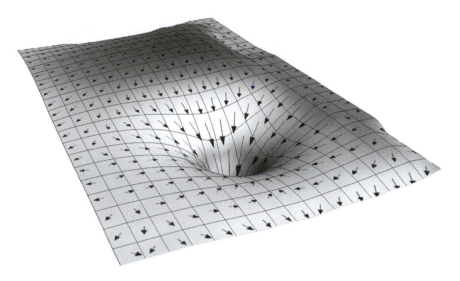

Figure 3.5. Water-table configuration showing a cone of depression forming around a pumping well. The magnitude of the hydraulic gradient is represented by the arrows increasing in length closer to the pumping well.

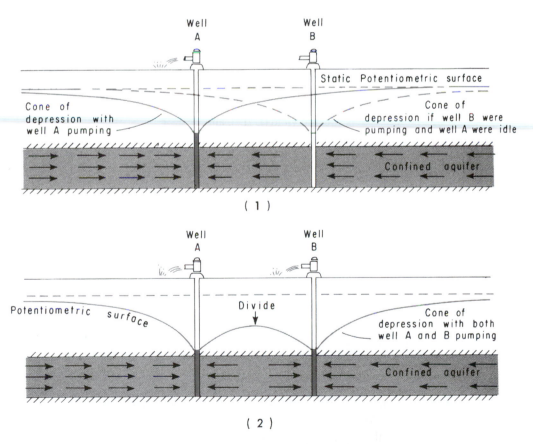

Figure 3.6. Well interference between two pumping wells (A and B) showing the additive nature of drawdown: (1) shows individual cones of depression around each well and (2) shows the composite cone of depression produced by both wells (modified from Heath, 1998).

The Theis equation can be used to compute the size and shape of a composite cone of depression from two or more pumping wells.   These calculations become more complicated if several wells are pumping, the wells are at different distances from the points of interest, and they are pumping at different rates.   The following general form of the Theis equation is used to make these composite drawdown calculations of the long-term behavior in an unconfined aquifer.

$$s_{total} = \frac{Q_1}{4\pi T}W(u_1) + \frac{Q_2}{4\pi T}W(u_2) + ... + \frac{Q_n}{4\pi T}W(u_n) \qquad (3\text{-}6)$$

where

s_{total}	is the composite drawdown at the point of interest produced by n pumping wells (L),
n	is the number of wells,
Q_n	is the discharge rate of each pumping well (L^3/T),
T	is the transmissivity (L^2/T), and
u_n	is the Theis equation parameter for each well.

$$u_n = \frac{r_n^2 S_y}{4Tt} \qquad (3\text{-}7)$$

where

r_n	is radial distance from each pumping well to the point of interest (L),
S_y	is specific yield (L^3/L^3), and
t	is elapsed time since the beginning of the pumping stress (T).

As you can no doubt appreciate, spreadsheets can greatly assist in making these calculations.

Excel Tips

- FACT(n) is an intrinsic *Excel* function that takes the factorial of the number "n." Use FACT(n) to increase the efficiency of programming the W(u) function.
- PI() inserts the constant π into an equation.
- LN(n) calculates the natural logarithm of the number "*n*."
- SUM() sums the cells designated within the parentheses.
- Cell anchor: $column$row anchors the cell (column, row) in a formula. Performing a relative copy will maintain the anchored cell within the formula and change only the other non-anchored cells in the formula (e.g., =$D$12+D2 will anchor cell D12 and not cell D2 during a relative copy).
- Exponential symbol: ^ raises a number to the designated power (e.g. 3^2 = 9).
- Other intrinsic *Excel* functions are listed and explained by clicking the f_x button in the Toolbar.

- Using the F2 key on a cell with a formula will highlight the reference cells used in the formula. This highlighting allows you to more easily evaluate the accuracy of the formula programmed into the cell.
- *Excel* will give an error message if a programming error occurs such as illustrated below with missing parentheses. However, *Excel* will not necessarily give an error if the formula is entered incorrectly.

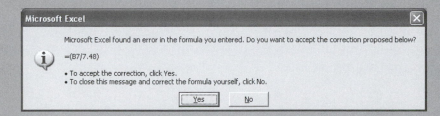

- "#DIV/0!" will appear in cells if a division by zero error has occurred. "#DIV/0!" displays if the denominator in any formula is zero, including formulas in reference cells that may also contain an error. Thus, one "#DIV/0!" error may propagate through the worksheet to multiple cells. Check your formulas for accuracy to avoid these errors.
- "#NUM" will appear in the $W(u_n)$ terms prior to programming the input parameters in the worksheet. This designation should resolve once the worksheet programming is complete.
- Save your work often to avoid losing information.

Parameter	Definition
b	Saturated thickness of the aquifer (L)
h_o	Initial hydraulic head at any distance (L)
h	Hydraulic head at the same distance after elapsed time (L)
n	Number of wells
Q_n	Discharge from each pumping well (L^3/T)
r_n	Radial distance from the pumping well to any point (L)
S	Drawdown at any time and distance from the pumping well (L)
S_{total}	Composite drawdown at the point of interest produced by n pumping wells (L)
S_y	Specific yield (L^3/L^3)
T	Transmissivity (L^2/T)
t	Elapsed time since the beginning of the pumping stress (T)
u_n	Theis equation parameter for each well
$W(u)$	Theis well function

Table 3.2 Parameter Definition Table – Chapter 3, Problem 2

CHAPTER 4
STREAM/AQUIFER INTERACTIONS

Principles & Concepts

Surface waters such as lakes and streams usually serve as areas of recharge and discharge for local and regional groundwater flow systems. The interactions between surface water and groundwater occur in a multitude of ways depending on hydrologic and geologic factors. In this chapter, we explore the factors that contribute to the exchange of water between surface water and aquifer systems. Three of the geologic factors effecting this exchange are the permeability of the streambed material and the transmissivity and coefficient of storage of the underlying aquifer. Other hydrologic factors are more dynamic over short time periods, such as stream stage (water elevation), water-level fluctuations caused by precipitation or drought, discharge to springs, and pumping of nearby wells. This chapter addresses some of the factors influencing the exchange of water between surface water and groundwater and some of the field techniques that are used to measure these exchanges.

Figure 4.1. Roaring Springs on the north rim of the Grand Canyon is an example of groundwater discharging at the surface to form a local stream (courtesy of Abe Springer).

Stream Gaging—Measuring the Exchange of Water

The rate of water exchange between an aquifer and a stream is commonly measured by determining the change in stream discharge at two locations along the stream. Stream gaging is used to obtain a "snapshot" of the discharge rate of a stream at a single profile across the stream; this discharge rate can be compared with discharge rates measured at other locations along the stream. The difference in the discharge rates at two locations may indicate the volumetric rate of water exchanged between the stream and aquifer.

Figure 4.2. Stream gaging using a wading rod, flow meter, and tag line to determine the rate of stream discharge at a single profile across the stream.

A gaining stream is one in which groundwater leaves an aquifer and enters a stream through discharge into the streambed or wetlands or a spring along the floodplain. A losing stream is one in which surface water enters an aquifer from the stream by infiltration across the streambed. The direction of water exchange is controlled by the difference in water levels between the aquifer and the stream.

A variety of factors influence the dynamics of water exchange between groundwater and surface water. Precipitation within the watershed may increase overland flow and cause higher stream stages, which change the hydraulic gradients between the stream and the aquifer. Floods usually force water into bank storage near the stream; the banks then slowly discharge this temporarily stored water as floodwaters recede. Subsequent drought may lead to a natural decline of the water table creating what is called baseflow recession, a prolonged decrease in the amount of groundwater discharging to the stream. (Baseflow discharge is the amount of water that enters the stream from an aquifer.)

Induced Infiltration

A stream may change between gaining and losing in response to human activities. Pumping wells located near surface-water bodies can have a pronounced effect. Figure 4.3A shows a simple flow system with no pumping well and groundwater discharging into the stream under the influence of the regional hydraulic gradient. In this case, the stream is considered gaining because groundwater is contributing to the flow of surface water. If a well is installed next to the stream and pumped at a low rate, a cone of depression forms that removes groundwater from storage and intercepts some groundwater flow that would normally discharge to the stream. This removal creates a local groundwater divide between the well and the stream as shown on Figure 4.3B. Groundwater discharges to the stream at a reduced rate because groundwater is captured by the well. This groundwater capture might be observed as a decrease in the streamflow downstream of the pumping well.

If the well pumps at a high rate, the cone of depression may extend to the stream causing a reversal of the local hydraulic gradient. In this case, groundwater is removed from aquifer storage and surface water is induced to flow across the streambed into the aquifer (induced infiltration). The well also captures groundwater flow that normally would discharge through the streambed. The discharge of the stream decreases downstream of the well because the well intercepts groundwater that otherwise would discharge to the stream and because the well induces infiltration from the stream into the aquifer. Measuring streamflow upstream and downstream of the pumping well location could show a decrease in streamflow.

Land-Use Modifications

Land use can also influence the amount of precipitation infiltrating into an aquifer. For example, urbanization increases the area of impervious surfaces (roadways, parking lots, rooftops, and other hard, poorly permeable surfaces), which modifies the hydrologic connectivity between infiltrating precipitation and recharge to the groundwater system. This interruption in the normal hydrologic cycle increases peak runoff to local streams, especially in watersheds with storm-water drainage systems. These land-use modifications can reduce baseflow discharge to a stream and change surface runoff characteristics (Leopold, 1968).

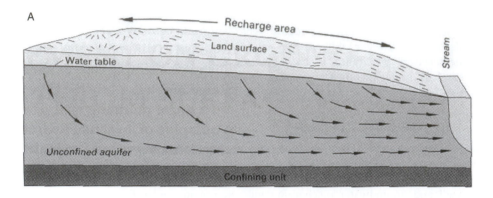

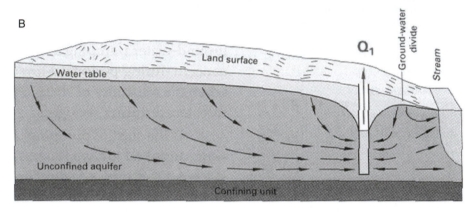

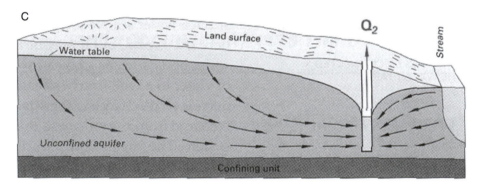

Figure 4.3. Illustrations showing the effect of a pumping well on a flow system. (A) Gaining stream where groundwater discharges into the stream. (B) Low rate of pumping and reversal of hydraulic gradient between the well and the stream, but no induced infiltration. (C) High rate of pumping causing a reversal of the hydraulic gradient and induced infiltration from the stream into the aquifer (from Winter et al., 1998).

Problem 1

GAGING STREAMFLOW GAIN and LOSS

Aberjona River, Woburn, Massachusetts

Overview

Measurement of the rate (volume per time) of surface water flowing through a cross-sectional area of a river is termed stream gaging. It represents one of the principal measures used by groundwater hydrologists to understand the movement of water throughout the watershed. Streamflow or stream discharge is usually calculated by multiplying the width of the stream by the average depth of the stream by the average velocity of the stream. As we discuss below, the irregular morphology of the streambed surface, variations in stream depth, and variability in flow velocity can lead to complexities in gaging a stream or a river.

Stream Gaging

Over the years, scientists have acquired a vast amount of experience improving the accuracy of stream-discharge measurements. The U.S. Geological Survey (USGS), which is the primary government agency assigned to quantify discharge in the nation's rivers, has developed standard techniques for a wide range of streambed morphologies and flow characteristics. Central to most of these techniques is the choice of a proper gaging location, measurement of water velocity, and streambed morphology. The location of the stream gaging site depends on the purpose of the streamflow measurements. If the measurements are part of a long-term regional inventory of water resources, then the following criteria should be observed (Rantz et al., 1982) (Figure 4.1).

1. Overall straight section of the river 300 feet upstream and downstream of the site.
2. Total flow is confined to one channel or thalweg at all stages of river flow. Avoid braided stream reaches.
3. Streambed shows little evidence of scour or fill and is devoid of aquatic plant growth.
4. Ideally, unchanging natural control such as bedrock to constrict low and high flows.
5. Pools are present upstream for extremely low flows to ensure recording stage at low flow.
6. Avoid the impacts of tributaries and tides.
7. Site is readily accessible at all stages.

The measurement of the streamflow velocity can be determined by a variety of techniques; however, the most common is the velocity-area method that uses a wading rod or suspended flow meter (Rantz et al., 1982). Many flow meters operate using a rotating propeller that spins at a rate proportional to the streamflow velocity. The number of revolutions per unit time is converted into a measure of streamflow velocity.

Rantz et al. (1982) describe a variety of flow meters used by the USGS. Recently, other methods have been developed to automate data collection. Common flow meters include the Price-Type AA used by the USGS on a wading rod or other suspension mechanism. This meter is designed for moderate stream velocities (0.25 to 8.0 feet/second). For smaller stream velocities (< 1 feet/second), the pygmy-scale flow meter is used, which is 40% of the scale of the Type AA meter (Rantz et al., 1982).

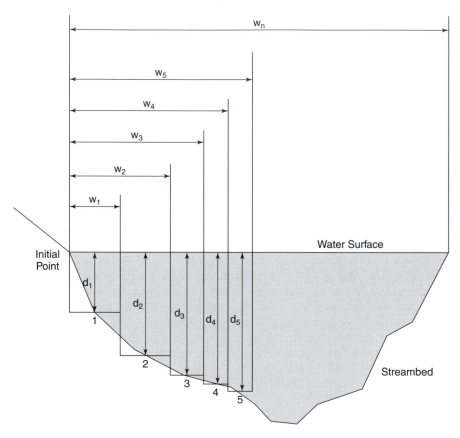

1, 2, 3, 4, 5, ..., n Compartments dividing stream cross section

w_1, w_2, w_3, w_4, w_5, ..., w_n Distance from initial point to observation poin.

d_1, d_2, d_3, d_4, d_5, ..., d_n Stream depth at observation point

Figure 4.4. The velocity distribution within a stream channel necessitates using compartmental divisions of the stream cross section for accurate gaging of stream discharge.

The stream cross section is usually divided into 20 to 30 compartments to approximate the irregular streambed morphology (Figure 4.4). The rule of thumb is that no more than 10% of the total flow should be within one compartment. Therefore, compartments may be of different widths and depths to ensure more accurate representation of velocity variations. Velocity measurements are taken at a depth of 60% from the water surface to the streambed for streams less than 2.5 feet deep. For deeper streams (> 2.5 feet), two velocity measurements are taken at 20% and 80% of the stream depth and the velocities are averaged to represent the overall velocity in the compartment on the basis of empirical evidence. The discharge from each compartment is calculated using the average velocity measurement and the cross-sectional area of the compartment. Then, all discharges from the compartments are summed to produce an average discharge along the entire stream cross section at the gaging location.

This method is a labor-intensive one that produces accurate results but determines the discharge only at a snapshot in time and at a single location. One can determine the continuous discharge of a stream by collecting continuous water surface elevation (stream stage) data. Stage data are collected by a variety of methods such as a stilling well used with float sensor or a bubble-gage and an analog or digital recorder of stage. The stage data are then compared with velocity-area discharge measurements at discrete flows (low, medium, and high). From this comparison, a stream rating curve is constructed to convert the continuous stream stage measurements to continuous discharge measurements at the given site. Rating curves need to be updated periodically because erosion and deposition in the streambed change flow velocities (Rantz et al., 1982).

Discharge Calculations

In this exercise, streamflow velocities were measured by a USGS researcher who used a pygmy-scale flow meter while wading across the Aberjona River in Woburn, Massachusetts, near the bridges on Olympia Avenue and Salem Street. Most flow meters have an empirical relation defined by the manufacturer between the rate of propeller rotation and the velocity of streamflow. The empirical relation of the current meter used at Woburn was

$$V = 0.977\,R + 0.028 \qquad (4\text{-}1)$$

where

V is the stream velocity in feet/second (L/T), and
R is the revolutions of the propeller per second (rotations/T).

The discharge of each compartment (Figure 4.1) is then calculated using the cross-sectional area of the compartment and the computed stream velocity in the compartment, such that

$$A = W\,D \qquad (4\text{-}2)$$

where

A is the cross-sectional area of the compartment (L^2),
W is the width of the compartment (L), and
D is the depth of the compartment (L).

The total stream discharge is determined by summing the discharges of the individual compartments, such that

$$Q = \sum_{i=1}^{m} V_i A_i \tag{4-3}$$

where

Q	is the total stream discharge (L^3/T),
V_i	is the stream velocity in the ith compartment (L/T),
A_i	is the cross-sectional area in the ith compartment (L^2), and
m	is the number of compartments in stream cross section.

See the Principles & Concepts section at the beginning of this chapter for additional description of induced infiltration as a result of groundwater pumping near a stream and the effect of infiltration and precipitation on streamflow.

Excel Tips

- SUM() sums the cells designated within the parentheses.
- Cell anchor: $column$row anchors the cell (column, row) in a formula. Performing a relative copy will maintain the anchored cell within the formula and change only the non-anchored cells (e.g., =$D$12+D2 will anchor cell D12 and not cell D2 during a relative copy).
- Exponent symbol: ^ raises a number to the designated power (e.g. 3^2 = 9).
- Other intrinsic *Excel* functions are listed and explained by clicking the f_x button in the Toolbar.

Parameter	*Definition*
A	Cross-sectional area (L^2)
D	Depth of stream compartment (L)
m	Number of compartments in stream cross section
Q	Total stream discharge (L^3/T)
R	Revolutions of propeller per second (Rotations/T)
V	Stream velocity (L/T)
W	Width of stream compartment (L)

Table 4.1 Parameter Definition Table – Chapter 4, Problem 1

Problem 2

FLOOD FREQUENCY AND URBANIZATION

Aberjona River Watershed, Massachusetts

Flood Frequency

Floods are natural phenomena that are important to the dynamic nature and health of rivers and streams. The disruptive and costly consequences of these natural events on society have resulted in environmental planning and risk assessment for populated areas near rivers and streams. In the United States, federal government agencies have been key to the development and use of various methods of determining flood frequency and flood probability. Some of these methods include the lognormal distribution (Weibull Method) and the Gumbel Type I, Gumbel Type III, and Log Pearson Type III distributions (Dunne and Leopold, 1978; Riggs, 1989). The publication *Guidelines For Determining Flood Flow Frequency* (U.S. Water Resources Council, 1981) describes the data and procedures for determination of flood frequency curves using various methods.

This exercise uses the Weibull Method (Ritter et al., 1995) to calculate flood recurrence intervals by taking the average time between two floods of equal or greater magnitude. The recurrence interval is calculated from a series of floods considered to be the peak stream discharges during a series of years. The recurrence interval is an expression of the probability of a flood of a given magnitude occurring in any one year, such that

$$T = \frac{n+1}{m} \qquad\qquad (4\text{-}4)$$

where

T	is the recurrence interval in years (T),
n	is the number of annual peak stream discharge values in the data series, and
m	is the magnitude rank: 1 is the largest discharge, 2 is the second-largest discharge,..., and m is the smallest discharge in the data series.

The results of the recurrence interval calculations for each annual peak discharge are plotted with recurrence interval on a logarithmic scale on the *x*-axis versus the peak discharge on an arithmetic scale on the *y*-axis (Figure 4.5). Hydrologists use these plots to forecast the probability of floods exceeding the period of record. Extrapolations of recurrence intervals are commonly made beyond the period of record to achieve a regulatory standard such as the 50-year and 100-year floods (Singh, 1987).

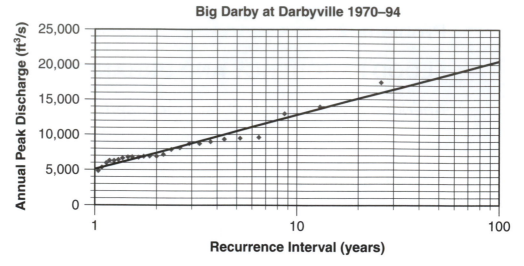

Figure 4.5. Extrapolated trendline used to predict magnitude of a 100-year flood based on annual peak streamflow data.

The exceedance probability of a flood discharge in any one year can be determined by

$$P = \frac{1}{T} \qquad (4\text{-}5)$$

where

P is the exceedance probability, and
T is the recurrence interval (T).

The exceedance probability is the likelihood of a flood of a given magnitude in any one year. For example, a flood with a recurrence interval of 100 years has an exceedance probability of 1% (1/100) in any one year (Riggs, 1987).

Urbanization

The amount of precipitation that becomes runoff is influenced by the rate of precipitation, soil type, soil moisture, slope of the land surface, vegetation, and land use in the watershed. As urban regions expand, the area of impervious surfaces commonly increases on or near the land surface, and runoff is commonly diverted quickly into retention ponds and storm sewers. These impervious surfaces occur as roadways, parking lots, rooftops, and other hard surfaces that intercept precipitation and usually provide controlled drainage through a storm water drainage system. Land usage can modify the hydrologic connectivity between atmospheric moisture and the unsaturated (vadose) and saturated (phreatic) zones by creating a short circuit in the hydrological cycle. Extensive urbanization can increase flood peak magnitudes because of the accelerated response of surface drainage and the reduced lag time between precipitation events and streamflow peaks. This relation is illustrated in Figure 4.6, where there is a higher peak flow and shorter lag time between the center of mass of the precipitation and the center of mass of the runoff in the urbanized case than in the non-urbanized case. Streams in urbanized areas usually experience less frequent low flows between storm events because less soil moisture is replenished and less groundwater is stored. Leopold (1968) documented that the increase in the average annual peak flow for many streams in urbanized areas is between 1 to 6 times greater depending on the amount of impervious surfaces and the amount of area drained by stormwater systems. As a consequence, streamflow discharge increases for a given recurrence interval as more of the watershed area is drained by stormwater drainage systems and covered by impervious surfaces.

Urbanization of a watershed also alters water quality and modifies stream hydrodynamics. Water commonly transports larger volumes of sediment, especially in areas of unprotected soil such as construction sites. Water quality commonly is degraded by increased sewage effluent from wastewater treatment plants and increased hydrocarbon contamination from surface-water runoff. Hydrodynamic modification occurs in urbanized watersheds because of the increase in runoff and peak flow magnitudes. These modifications increase channel instability, increase scour, decrease riparian vegetation, and enlarge the channel.

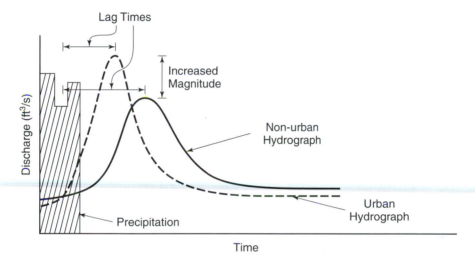

Figure 4.6. Storm hydrograph for an urbanized and a non-urbanized watershed given similar precipitation amounts. Note the increase in magnitude of peak flow and shortened lag time between the precipitation event and the peak flow of the stream. Less frequent low flows occur between storm events in an urbanized watershed.

Baseflow

In many humid areas, groundwater discharges into streams by seeping into the streambed. Groundwater discharge into a stream is termed baseflow. The amount of baseflow entering a stream is directly proportional to the hydraulic gradient driving the groundwater into the stream. The amount of baseflow can be estimated from storm hydrographs.

Figure 4.7 illustrates the hydrologic cycle and, in particular, baseflow discharge to a stream. In the absence of impervious surfaces, a portion of the precipitation infiltrates through the unsaturated zone and reaches the groundwater flow system at the water table. Locally, this infiltrated water causes the water table to rise at the point of recharge, which increases the hydraulic gradient toward the discharge location. Because baseflow is directly proportional to hydraulic gradient, the amount of baseflow entering the stream increases as the hydraulic gradient to the stream increases.

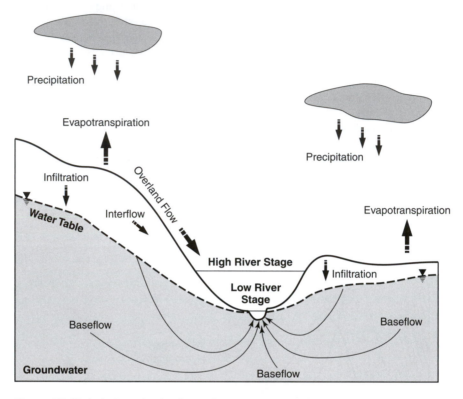

Figure 4.7. Hydrologic cycle showing major components and including baseflow discharge to streams.

Infiltration becomes limited as the area of impervious surfaces increases in the watershed. Limited infiltration leads to a limited response of the hydraulic gradient to a storm event and subsequently to diminished baseflow discharging into the stream. Thus, as urbanization proceeds, a greater percentage of streamflow is derived from overland flow and engineered drainage and a lower percentage is contributed by baseflow from the groundwater flow system.

The baseflow component of total streamflow can be determined by a variety of methods, one of which is the analysis of several years of streamflow records. Rutledge (1998) developed an algorithm for separating the baseflow component from total streamflow. We applied this algorithm to the streamflow records used in this exercise. This program estimates daily baseflow values based on a fit to antecedent recessions and then interpolates the groundwater discharge at other times in the streamflow record (Rutledge, 1998).

Excel Tips

- SUM() computes the sum of the values in cells inside the parentheses.
- STDEV() is an intrinsic function in *Excel* and computes the standard deviation of the values in the cells inside the parentheses.
- Sorting data: To assign the magnitude rank to each annual peak stream discharge value, sort the data from largest to smallest (descending order). Sorting can be performed in *Excel* by selecting the data you want to sort (everything except the header information) and choosing sort data (Data >> Sort). Specifically, select the column with the annual peak discharge values and identify "Descending." These choices will sort the data for easier assignment of the magnitude rank (see below).

- Plotting graphs: Highlight the data you wish to plot and choose the Chart Wizard (see below). Work your way through the selections in the four dialog boxes (chart type, source data, options, and location) that assist you in creating the appropriate plot. Be sure to properly label the axes of the graphs and include the units in the label. We suggest creating the plot on a separate worksheet.

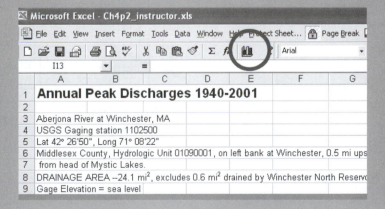

- Logarithmic scale: The identified axis of a plot can be converted to a logarithmic scale by double clicking on the desired axis, which opens the "Format Axis" dialog box. Near the bottom of the box, check the "Logarithmic scale" box.

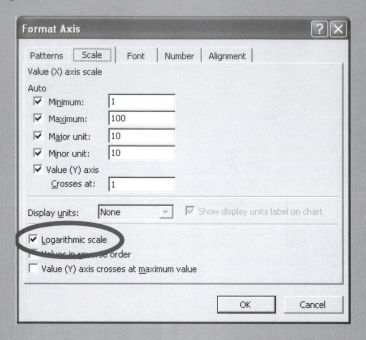

- Adding gridlines: Adding vertical gridlines to the recurrence interval graph can assist you in determining the recurrence intervals of various flood magnitudes. Under the "Chart Options" dialog box, choose "Gridlines" and make sure both major and minor gridlines are drawn for the x-axis.

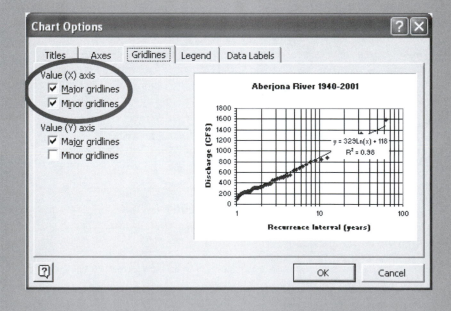

- Adding trendlines: Trendlines or best-fit lines can be added to an *Excel* plot by right mouse clicks when the cursor is positioned over the data points. Clicking produces the dialog box used to add a trendline. For the flood frequency analysis, add a straight line through the plot of recurrence intervals versus annual peak discharges.   If the recurrence interval axis is on a logarithmic scale, the chosen trendline must also be logarithmic for it to appear as a straight line on the plot (see below). The user can also add the equation of the trendline and the R^2 value, which is useful in answering the questions. The R^2 value is also known as the coefficient of determination, a measure of how well the trendline fits the actual data. Values of R^2 range from 0 to 1; 1 indicates the most reliable trendline.

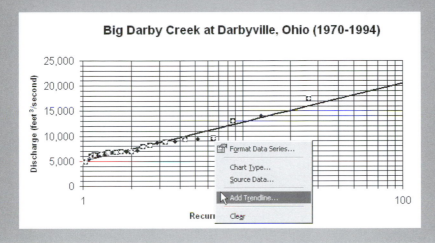

Parameter	Definition
m	Magnitude rank
n	Number of years of record
P	Exceedance probability (1/T)
T	Recurrence interval (T)

Table 4.2 Parameter Definition Table – Chapter 4, Problem 2

CHAPTER 5

AQUIFER TESTING TECHNIQUES

Principles & Concepts

Much of the quantitative development of groundwater hydrology can be traced back to the work of Henry Darcy (1856) who, as a professional engineer, published a famous report on the water supply for his home city of Dijon, France. The results of his experiments with the flow of water through a sand-filled column became the basis for Darcy's Law, which can be stated as

$$Q = -K A \left(\frac{dh}{dL} \right) \tag{5-1}$$

where

Q	is discharge (L^3/T),
K	is hydraulic conductivity (L/T),
A	is cross-sectional area through which flow occurs (L^2), and
$\left(\dfrac{dh}{dL} \right)$	is hydraulic gradient (L/L).

The hydraulic conductivity term in Darcy's Law describes the ability of a porous medium to transmit a fluid; that ability is a function of the properties of the porous medium and the properties of the fluid moving through it. One can separate the properties of the porous medium from the properties of the fluid using the following equation.

$$K = \frac{k \rho g}{\mu} \tag{5-2}$$

where

k	is the intrinsic permeability of the porous medium (L^2),
ρ	is the fluid density (M/L^3),
g	is the acceleration of gravity (L/T^2), and
μ	is the dynamic viscosity of the fluid (M/LT).

Intrinsic permeability is usually used in the petroleum industry because the movement of oil, gas, and water requires evaluation of variable fluid density and multiple fluid phases. The units of intrinsic permeability are often measured in centimeters squared and meters squared or darcys (1 darcy = 10^{-8} cm^2).

Intrinsic permeability is also used when addressing problems related to saltwater intrusion, brine migration, and injection of fluid wastes.

This chapter describes several techniques used to quantify the physical parameters of an aquifer system. In addition to hydraulic conductivity, these parameters include porosity (n), compressibility of the aquifer skeleton (α), compressibility of water (β), transmissivity (T), specific storage (S_s), coefficient of storage (S), and specific yield (S_y). In the past 100 years, considerable research has been completed to develop laboratory methods and field techniques to quantify each of these physical parameters in different types of geologic materials and hydrologic settings (Kruseman and de Ridder, 1990).

Laboratory Techniques

Close examination of the porous media that compose aquifers indicates that water flows around the individual sediment or mineral grains in interconnected void spaces (Figure 5.1). Hazen (1911) developed an empirical relation between the hydraulic conductivity and the grain-size distribution of unconsolidated sand. This first approximation of the hydraulic conductivity is important in many studies where more extensive determinations of hydraulic parameters are not available. Grain-size distribution can be determined by a variety of methods, such as sieve analysis for sand-sized particles, as discussed in Problem 1 of this chapter.

Figure 5.1. Microscopic view of a thin section of St. Peter Sandstone showing quartz grains (pale gray) and primary porosity (dark epoxy) (courtesy of Dave Houseknecht, USGS Reston).

Field Techniques

Central to many of the other techniques are field observations of water-level changes in response to an imposed hydraulic stress. For example, removing water from a well and measuring the temporal and spatial response of hydraulic head in the aquifer can be used to determine site-specific values of some of these physical parameters.   Theoretical models of aquifer behavior have been developed to describe the stress response in a variety of geologic settings. The measured response in the aquifer is compared with the theoretical model to determine values of the physical parameters.

Figure 5.2. A water-filled PVC slug was lowered down the test well to produce an "instantaneous" hydraulic response recorded by the pressure transducer connected to the data logger (cord going down the well).

Slug tests use a device (a slug) to instantaneously disturb water levels in a well and in the halo of the aquifer immediately adjacent to it (Problems 2 and 3). Commonly, the slug consists of a closed polyvinyl chloride (PVC) tube filled with a mixture of water and sand to prevent its buoyancy in the well. Other types of slugs include a 55-gallon drum filled with water that is quickly released into the test well to raise the water level. Pneumatic slugs can also be used: the water level is initially depressed by injecting compressed air into the casing above the water level, then the compressed air is instantaneously released from the casing and the water level rises to its equilibrium level.

The aquifer response to the imposed hydraulic stress (i.e., the slug) is measured as the water-level response in the tested well. The water-level response data then can be analyzed according to characteristics of the borehole, aquifer system, and test configurations to determine hydraulic parameters such as hydraulic conductivity (Kruseman and de Ridder, 1990; Hyder et al., 1994).

Figure 5.3. Various sizes of PVC slugs used to perturb the water level
in a well to initiate a slug test.

Hvorslev (1951) developed a method for analyzing slug test water-level
response data in confined aquifers assuming various aquifer and hydraulic
characteristics discussed in Problem 2 of this chapter.   Since Hvorslev's
development of methods for analyzing slug tests (Problem 2), hydrogeologists
and petroleum geologists have reinvented and refined many different types of
variable-head, single-well tests known as slug tests, bailer tests, and rising or
falling head tests. Slug tests have been widely used because they provide values
for hydraulic parameters rapidly, require simple logistical investment, and are
low cost. Slug tests do not discharge or withdraw contaminated water, which is
an additional advantage in many site characterization studies (Chirlin, 1990).
Subsequent slug test methods, such as those of Cooper et al. (1967), Moench and
Hsieh (1985), Butler and McElwee (1990) and Hyder et al. (1994), remove some
simplifying assumptions such as a fully penetrating well and accounting for well
skin characteristics in confined aquifers. The Bouwer and Rice (1976) slug test
method (Problem 3 in this chapter) was developed for analysis of unconfined
aquifers but has been successfully applied to confined aquifers (Bouwer, 1989).

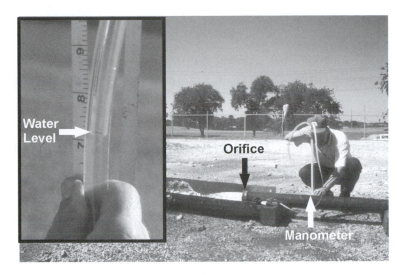

Figure 5.4. Measurement of pumping well discharge during an aquifer test using an orifice and manometer. Inset shows a close up of manometer reading on a graduated scale.

Slug tests have limited hydraulic effect on the tested aquifer; therefore, they produce estimates of hydraulic parameters only within a limited area around the test well. Aquifer tests withdraw large volumes of water from a pumping well to impose a substantial hydraulic stress on the aquifer (Figure 5.4) (Problems 4 and 5 in this chapter). The zone of influence of an aquifer test is directly related to the rate of pumping and can be measured in the pumping well itself or adjacent observation wells completed within the aquifer (Figure 5.5). Design and completion of aquifer tests require substantial preparation and knowledge of the local geology and hydrogeology (Sanders, 1998; Fetter, 2001). The values of aquifer parameters, such as transmissivity and storage coefficient, derived from aquifer tests usually represent averages over the zone of influence from the pumping well. Determining which aquifer testing technique to use depends on the geology and hydrogeologic setting, the scope of the project, and other budgetary and logistic constraints.

Figure 5.5. Measurement of water levels in a well using an electric tape during an aquifer testing demonstration.

Problem 1

ESTIMATING HYDRAULIC CONDUCTIVITY USING THE HAZEN GRAIN-SIZE METHOD

Remediation of TCE and PCE at North Canton, Ohio

Overview

Grain-size distributions are used for a variety of purposes in the geosciences including the estimation of hydraulic conductivity of sand-sized sediments. Various techniques are used to determine grain-size distributions, such as sieve analysis, hydrometer analysis, and laser diffraction. Sieve analysis uses a series of decreasing-sized wire-mesh sieves to fractionate a sediment sample and then computes the weight percentages of particles passing through each sieve size. This analysis is limited to sand and gravel-size fractions (ASTM-D422, 2002) because of limitations in the fineness of sieve screens. Smaller particle sizes in the silt- to clay-sized fraction are measured using hydrometer, pipette, or laser diffraction analysis.

Sediment is classified into four main size categories: gravel, sand, silt, and clay. According to the Udden-Wentworth system, gravel is greater than 2 mm, sand is from 2 to 0.062 mm, silt is from 0.062 to 0.002 mm, and clay is smaller than 0.002 mm (Figure 5.6). The results of a particle size analysis are plotted on semi-logarithmic graph of grain size versus cumulative percent finer by weight (cumulative percent of the sediment weight that is finer than the given size relative to the total sample weight). The curve of grain-size distributions is used to determine the particle size characteristics of the sample and to compare samples. One important parameter computed on the basis of grain size is the uniformity coefficient, which is a measure of the degree of sorting (or, inversely, the degree of grading) of the sediment. It is the ratio of two grain sizes defined by the following formula.

$$C_u = \frac{d_{60}}{d_{10}} \tag{5-3}$$

where

C_u is the uniformity coefficient,
d_{60} is the grain size (e.g., mm) at 60% finer by weight (L), and
d_{10} is the grain size (e.g., mm) at 10% finer by weight (L).

Sediments with C_u values more than 6 are considered poorly sorted (well graded), whereas sediments with C_u values less than 4 are considered well sorted (poorly graded). See Figure 5.6 as an example of how to determine d_{60} and d_{10} from a grain-size distribution graph.

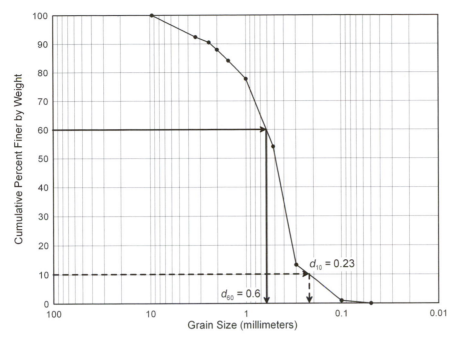

Figure 5.6. Example grain-size distribution curve showing d_{60} and d_{10} values derived from the graph.

Hazen (1911) used grain-size distributions in unconsolidated sands to estimate hydraulic conductivity. His work has proven useful as an estimation method where direct in-situ permeability measurements are sparse. Hazen's work showed that for sands with an effective grain size (d_{10}) of 0.1 to 3.0 mm, his empirical equation is a useful estimator of hydraulic conductivity. This relation depends on a power-law relation with hydraulic conductivity and has been shown to be applicable in sediment sizes ranging from fine sand to gravel.

$$K = C(d_{10})^2 \qquad\qquad (5\text{-}4)$$

where

K	is the hydraulic conductivity (cm/s),
d_{10}	is the effective grain size (cm) at 10% finer by weight, and
C	is the Hazen coefficient (1/cm-s) (see Table 5.1).

Hazen Coefficient (C)	*Sediment Texture*
40 to 80	Very fine sand, poorly sorted sand
40 to 80	Fine sand with appreciable fines
80 to 120	Medium sand, well sorted
80 to 120	Coarse sand, poorly sorted
120 to 150	Coarse sand, well sorted, clean

Table 5.1. Hazen Coefficients

Heterogeneity of Hydraulic Conductivity

The spatial distribution and directional characteristics of hydraulic conductivity play an important role in understanding the movement of groundwater. The Hazen method assumes that the porous media is isotropic (properties values are the same in all directions) with respect to hydraulic conductivity rather than anisotropic (different depending on direction). However, sedimentologists know from examination of sand deposits that texture and grain packing are important characteristics that affect values of porosity and hydraulic conductivity. Hazen's method assumes that the porous medium is isotropic. A porous medium that contains pathways of preferential flow would be considered anisotropic and not suitable for evaluation by Hazen's method.

Heterogeneities exist in most aquifers where the physical parameters such as permeability, porosity, storage, or thickness differ from place to place within the geologic unit or units composing the aquifer. In a layered system, where hydraulic conductivity differs in different layers within an aquifer, one can describe the average horizontal hydraulic conductivity using a weighted average. The hydraulic conductivity is weighted using the thickness of each layer (b_m).

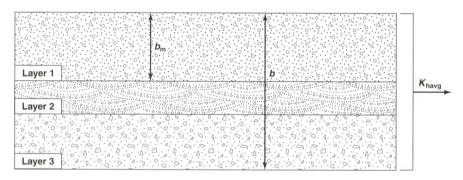

Figure 5.7. Hypothetical heterogeneous aquifer consisting of three distinct layers.

The average horizontal hydraulic conductivity in a layered system is determined with the following equation.

$$K_{havg} = \sum_{m=1}^{n} \frac{K_m b_m}{b} \qquad (5\text{-}5)$$

where

K_{havg} is the average horizontal hydraulic conductivity (L/T),
K_m is the hydraulic conductivity of the mth layer (L/T),
b_m is the thickness of the mth layer (L),
b is the total thickness of the layered system (L), and
n is the number of layers.

Figure 5.7 illustrates an example of a layered heterogeneous aquifer with three distinct layers. Transmissivity is a parameter that describes the ability of the entire thickness of the aquifer to transmit water (see Chapter 3). The average transmissivity of a layered system is calculated with the following equation.

$$T_{avg} = \frac{\sum_{m=1}^{n} K_m b_m}{n} \qquad (5\text{-}6)$$

where

T_{avg} is the average transmissivity (L^2/T),

K_m is the horizontal hydraulic conductivity of the mth layer (L/T),

b_m is the thickness of the mth layer (L), and

n is the number of layers.

Excel Tips

- SUM() sums the cells designated within the parentheses.
- Cell anchor: \$column\$row anchors the cell (column, row) in a formula. Performing a relative copy will maintain the anchored cell within the formula and change only the other non-anchored cells in the formula (e.g., =\$D\$12+D2 will anchor cell D12 and not cell D2 during a relative copy).
- Exponent symbol: ^ raises a number to the designated power (e.g., $3^2 = 9$).
- Other intrinsic *Excel* functions are listed and explained by clicking the f_x button.

Parameter	Definition
b	Thickness of the aquifer (L)
b_m	Thickness of the aquifer m^{th} layer (L)
C	Hazen coefficient (1/L-T)
C_u	Uniformity coefficient
d_{10}	Effective grain size at 10% finer by weight (L)
d_{60}	Effective grain size at 60% finer by weight (L)
K	Hydraulic conductivity (L/T)
K_{havg}	Average horizontal hydraulic conductivity (L/T)
K_m	Horizontal hydraulic conductivity of the mth layer (L/T)
n	Number of layers
T	Transmissivity (L^2/T)
T_{avg}	Average transmissivity (L^2/T)

Table 5.2 Parameter Definition Table – Chapter 5, Problem 1

Problem 2

ESTIMATING HYDRAULIC CONDUCTIVITY USING THE HVORSLEV SLUG-TEST METHOD

Slate Ridge Estates Development, Fairfield County, Ohio

Overview

M. Juul Hvorslev was an engineer who worked in the U.S. Army Corps of Engineers at the Waterways Experiment Station in Mississippi. In his work to understand water movement in soils for water supply and proper design of foundations, he developed and published one of the earliest slug-test methods for estimating values of hydraulic conductivity in confined aquifers. This analytical method requires several simplifying assumptions, including homogeneous, isotropic, and uniform thickness of the aquifer, potentiometric surface that is nearly horizontal throughout the test-influence area, fully penetrating well, negligible specific storage, and a finite effective radius. The Hvorslev (1951) method also can be used in unconfined aquifers if the water table is sufficiently above the top of the well screen. This method is not appropriate in low hydraulic conductivity well-skin conditions because it yields estimates that are heavily weighted toward the hydraulic conductivity of the well skin (well skin is the area immediately adjacent to the wellbore, which may have different hydraulic properties as a result of the drilling process than the native, undisturbed aquifer material).

The estimated value of hydraulic conductivity depends on determination of the effective well-screen length and radius. The effective well-screen length and radius include the sand pack (if present) in the annular space around the well (see Figure 5.8). Thus, the Hvorslev method requires information on well construction to provide the effective screen length and radius.

Hvorslev (1951) discusses many different geometric configurations for slug-tested wells. We present the formulation that requires the effective length of the well screen to be eight times the effective radius of the well screen ($L_e/R > 8$), which is a common design for monitoring wells. Using this formulation, Hvorslev (1951) developed the following equation to calculate hydraulic conductivity.

$$K = \frac{r^2 \ln\left(\dfrac{L_e}{R}\right)}{2 L_e T_o} \tag{5-7}$$

Ground Surface

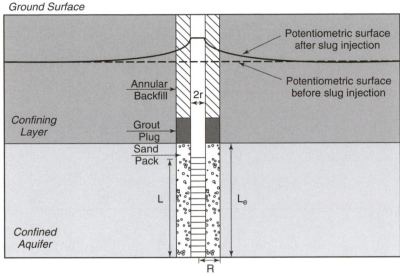

Figure 5.8. Well construction parameters needed for slug-test analysis by the Hvorslev method in a confined aquifer.

where

K is hydraulic conductivity of the aquifer (L/T),

r is the radius of the casing (L),

R is the effective radius of well screen (L),

L_e is the effective length of the well screen (L), and

T_0 is the time lag required for the potentiometric surface to rise or fall to 37% of the initial water-level change (T).

Values of r, R, and L_e are based on the geometry and construction of the well. T_0 is based on the measured response data (i.e., change in water level with time) when the slug is injected or withdrawn. The value of T_0 is determined by plotting the logarithm of the normalized head (h/h_0) versus time since the beginning of the test (Figure 5.9), where h_0 is the water level in the well immediately following the injection or withdrawal of the slug and h represents subsequent measured values of hydraulic head. Each value of h is normalized to the initial head change (h_0); thus, values of the normalized head (h/h_0) are calculated. During the test, the normalized head declines from 1.0 back to zero as the water level recovers back to its equilibrium position. Next, a straight line is fitted through these data points to approximate the behavior of the aquifer to the induced hydraulic stress produced by the slug. T_0 is the time lag (x-axis value) when the fitted line crosses the h/h_0 ratio of 0.37 (i.e., 37% of the initial change).

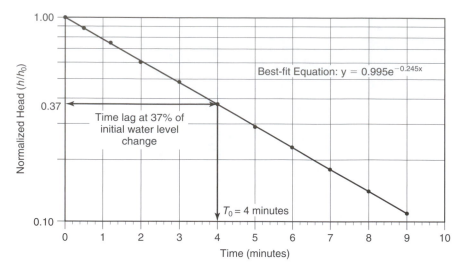

Figure 5.9. Example of Hvorslev method plot of head ratio (h/h_0) versus time to determine T_0.

When the response data have a pronounced concave-upward curvature as in Figure 5.10, the aquifer is exhibiting non-negligible storage. Fitting a line through these data depends on the situation. Butler (1998) suggests that the best-fit line to the normalized response data is in the range of h/h_0 equal to 0.15 to 0.25. This suggestion is similar to that of Bouwer (1989) to account for the double-straight line response.

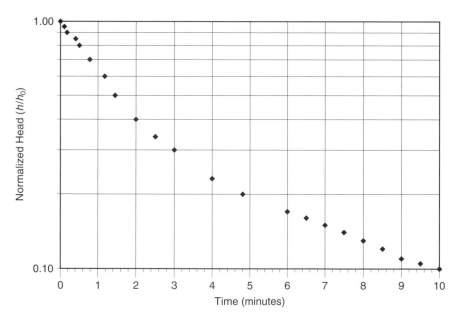

Figure 5.10. Concave-upward curvature of response data indicates that aquifer storage is not negligible.

Commonly, the change in the normalized head values (h/h_0) versus time appears to be linear on the semi-logarithmic plot; however, in some analyses the initial response may appear displaced relative to the other head values. This offset of the initial data points from a straight line may be the result of well-skin effects or other hydrologic interferences. In this case, Hvorslev (1951) suggests a correction to the fitted line through the data points. This correction entails fitting a straight line to the majority of the available data while ignoring the initial head-change values at the beginning of the slug test. This best-fit line is then offset (i.e., moved vertically) until it passes through the values of 1.0 for h/h_0 at time equal to zero (Figure 5.11). This offset correction procedure is necessary in the exercise presented. The value of T_0 needed for the Hvorslev method can then be determined by using the offset line.

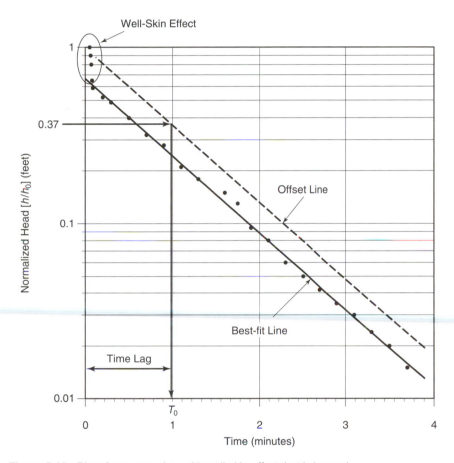

Figure 5.11. Plot of response data with well skin effect that is ignored in determining the best-fit line. The offset line is parallel to the best-fit line and intersects h/h_0 of 1.0 to determine the time lag (T_0) for Hvorslev analysis.

Excel Tips

- Cell anchor: $column$row anchors the cell (column, row) in a formula. Performing a relative copy will maintain the anchored cell within the formula and change only the other non-anchored cells in the formula (e.g.,=$D$12+D2 will anchor cell D12 and not cell D2 during a relative copy).
- Exponent symbol: ^ raises a number to the designated power (e.g., 3^2 = 9).
- LN(n) calculates the natural logarithm of the number "*n*."

- *Adding Trendlines* – Trendlines or best-fit lines can be added to *Excel* plots by right mouse clicks when the cursor is positioned over the data points. These clicks produce a dialog box that is used to add a trendline. If the *y*-axis is on a logarithmic scale, the trendline must be exponential for it to appear as a straight line on the plot. The user can also specify to add the equation of the trendline as well as the R^2 value, which is useful in answering the questions. The R^2 value is also known as the coefficient of determination, which measures how well the trendline fits the actual data. Values of R^2 range from 0 to 1; 1 indicates the most reliable trendline.

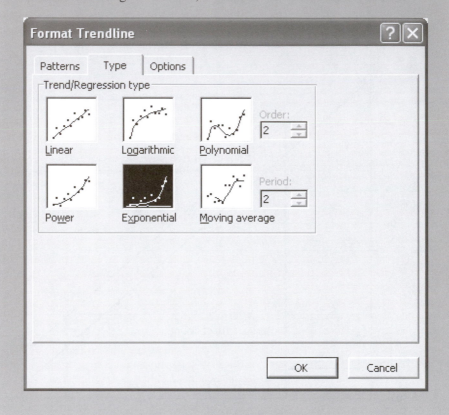

Parameter	Definition
K	Hydraulic conductivity of aquifer (L/T)
L	Length of well screen (L)
L_e	Effective length of the well screen (L)
r	Radius of the casing (L)
R	Effective radius of well screen (L)
T_0	Time required for the water level to rise or fall to 37% of the initial water-level change (T)

Table 5.3 Parameter Definition Table – Chapter 5, Problem 2

Problem 3

ESTIMATING HYDRAULIC CONDUCTIVITY USING THE BOUWER AND RICE SLUG-TEST METHOD

Freedom Street Well Field, North Canton, Ohio

Overview

Another useful method of slug-test analysis was developed by Bouwer and Rice (1976). This versatile method can be used in open boreholes and screened wells alike and is applicable to both fully and partially penetrating wells in unconfined aquifers with water levels above the screened interval. It can be used in a confined aquifer if the well screen is appreciably below the bottom of the upper confining layer. Similar to other slug-test methods, the Bouwer and Rice method assumes homogeneous, isotropic, and uniform thickness of the aquifer, a water table that is nearly horizontal throughout the test influence area, no borehole storage, negligible specific storage within the aquifer, and negligible well losses. In addition, the saturated thickness is assumed not to change during the test (Kruseman and de Ridder, 1990; Hyder et al., 1994; Butler, 1998).

The equation presented in Bouwer and Rice (1976) appears below.

$$K = \frac{r^2 \ln\left(\frac{R_e}{R}\right)}{2\,L_e}\;\frac{1}{t}\;\ln\frac{H_0}{H_t} \qquad (5\text{-}8)$$

where

K	is the hydraulic conductivity of the aquifer (L/T),
r	is the radius of well casing (L),
R	is the effective radius of the well screen (L),
R_e	is the effective radial distance through which the change in head is dissipated (L),
L_e	is the effective length of well screen or open section of the well (L),
H_0	is the hydraulic head at initial time (t = 0) (L),
H_t	is the hydraulic head measured in test well (t = t) (L), and
t	is the time since injection or withdrawal of slug (T).

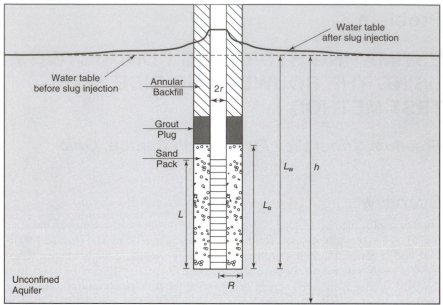

Figure 5.12. Well construction parameters needed for slug-test analysis using the Bouwer and Rice method.

The values for many of the parameters used in equation 5-8 can be obtained from well construction information (Figure 5.12) and from the hydraulic response data. However, the effective radial distance through which hydraulic conductivity is measured (R_e) must be derived from empirical values. Bouwer and Rice (1976) provide a methodology that uses well construction geometry and empirically derived values of the parameters A, B, and C (Figure 5.13) for determination of ln (R_e/R). These values are used to approximate a value of hydraulic conductivity and effective radial distance of the test in the *Excel* problem.

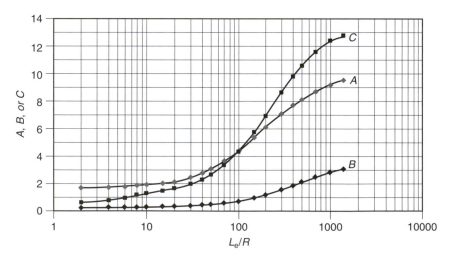

Figure 5.13. Polynomial approximation of A, B, and C curves for the Bouwer and Rice (1976) slug-test method.

For fully penetrating wells where L_w is equal to the saturated thickness (h), ln (R_e/R) has the following relation.

$$\ln \frac{R_e}{R} = \left[\frac{1.1}{\ln \left(\frac{L_w}{R} \right)} + \frac{C}{\left(\frac{L_e}{R} \right)} \right]^{-1} \qquad (5\text{-}9)$$

For partially penetrating wells where the length from the water table to the bottom of the well screen (L_w) is less than the saturated thickness (h), ln (R_e/R) has the following relation.

$$\ln \frac{R_e}{R} = \left[\frac{1.1}{\ln \left(\frac{L_w}{R} \right)} + \frac{A + B \ln \left[(h - L_w)/R \right]}{\left(\frac{L_e}{R} \right)} \right]^{-1} \qquad (5\text{-}10)$$

The last set of terms in the Bouwer and Rice equation 5-8 concerning time and hydraulic head is determined from the measured aquifer response. The change in hydraulic head from the initial water level is plotted versus time as the water levels re-equilibrate after injection or withdrawal of the slug. An example of this plot appears in Figure 5.14 (see next page).

The hydraulic head values generally approximate a straight line on a semi-logarithmic plot, and the value of $(1/t) \ln (H_0/H_t)$ can be determined from the graph using equation 5-11. The slope of the line is an expression of the hydraulic conductivity and is computed using any two points on the best-fit line. One point on the best-fit line lies at the intersection of H_1 and t_1 and the other point lies at the intersection of H_2 and t_2. These pairs are related in the following equation.

$$\frac{1}{t} \ln \left(\frac{H_0}{H_t} \right) = \left(\frac{1}{t_2 - t_1} \right) \ln \left(\frac{H_1}{H_2} \right) \qquad (5\text{-}11)$$

One complication of this method of analysis is that the field data may display two distinct line segments on the graph of change in hydraulic head versus time (Bouwer, 1989). The first response, similar to that seen on Figure 5.10, may be from the gravel pack around the test well and the second (later time) response from the native aquifer material. Thus, the interpreter must identify which line segment represents the aquifer response and use that response in the Bouwer and Rice method.

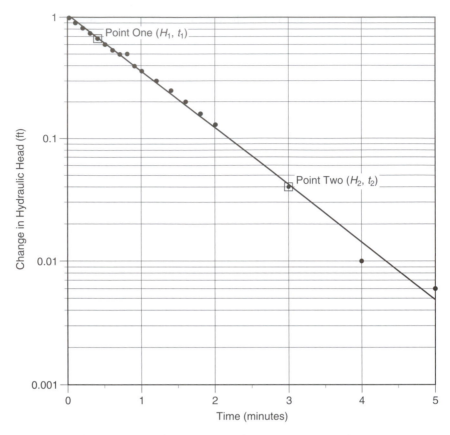

Figure 5.14. Example plot of the water-level response measured during a slug test when analyzed using the Bouwer and Rice method.

Excel Tips

- Cell anchor: $column$row anchors the cell (column, row) in a formula. Performing a relative copy will maintain the anchored cell within the formula and change only the other non-anchored cells in the formula (e.g., =$D$12+D2 will anchor cell D12 and not cell D2 during a relative copy).
- Exponent symbol: ^ raises a number to the designated power (e.g., $3{\wedge}2 = 9$).
- LN(n) calculates the natural logarithm of the number "*n*."
- LOG(n) calculates the logarithm base 10 of the number "*n*."
- EXP(n) calculates the exponential value of the number "*n*."
- PI() returns the values of π.
- *Adding trendlines* – Trendlines or best-fit lines can be added to *Excel* plots by right mouse clicks when the cursor is positioned over the data points. These clicks produce a dialog box that is used to add a trendline. If the *y*-axis is on a logarithmic scale, the trendline must be exponential for it to appear as a straight line on the plot. The user can also specify to add the equation of the trendline as well as the R^2 value, which is useful in answering the questions. The R^2 value is also known as the coefficient of determination, which measures how well the trendline fits the actual data. Values of R^2 range from 0 to 1; 1 indicates the most reliable trendline.

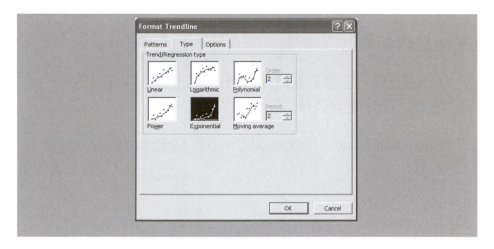

Parameter	Definition
A	Bouwer and Rice parameter
B	Bouwer and Rice parameter
C	Bouwer and Rice parameter
H_0	Hydraulic head at initial time ($t = 0$) (L)
H_t	Hydraulic head measured in the test well ($t = $ t) (L)
h	Saturated thickness of the aquifer (L)
K	Hydraulic conductivity of aquifer (L/T)
L_e	Effective length of the well screen (L)
L_w	Length from water table to bottom of well screen (L)
r	Radius of the casing (L)
R	Effective radius of well screen (L)
R_e	Effective radial distance that head is dissipated (L)
t	Time since injection or withdraw of slug (T)

Table 5.4 Parameter Definition Table – Chapter 5, Problem 3

Problem 4

AQUIFER TEST ANALYSIS USING THE THEIS TYPE-CURVE METHOD

Englewood Well Field, Southwest Sarasota County, Florida

Background

Theis (1935) used his mathematical description of the transient behavior of radial groundwater flow to a pumping well to calculate aquifer parameters such as transmissivity, hydraulic conductivity, and coefficient of storage. In this analysis, a pumping well within an aquifer creates a radial stress that depends on the rate of pumping, the geometry of the aquifer, and values of various aquifer parameters. By observing the response of water-level changes in the aquifer within the area influenced by pumping, Theis showed how to calculate site-specific values of aquifer parameters using a type-curve matching method (Fetter, 2001). The Theis method of analysis assumes transient flow within the aquifer; thus, water levels change with time of pumping and the aquifer storage coefficient is taken into account. The Theis type-curve method also requires the same assumptions described in Chapter 3 for a confined aquifer (isotropic, homogeneous, flat lying, infinite, overlain and underlain by an impermeable layer, and releases water instantaneously from storage) and for the pumping well (fully penetrates the aquifer, discharges at a constant rate, and lacks borehole storage).

To calculate values of these aquifer parameters, the Theis equation presented in equation 3-4 is rewritten in terms of the transmissivity of the aquifer. Thus, transmissivity can be expressed as a function of the pumping rate, the Theis well function, and the observed drawdown of the potentiometric surface in an observation well as expressed in the following equation.

$$T = \frac{Q}{4\pi(h_0 - h)} W(u) \qquad (5\text{-}12)$$

where

T is transmissivity of the aquifer (L^2/T),
Q is constant discharge from pumping well (L^3/T),
h is hydraulic head after elapsed time (L),
h_0 is initial hydraulic head (L), and
$W(u)$ is the Theis well function.

The storage coefficient of the aquifer can be determined by solving equation 3-2 for the storage coefficient.

$$S = \frac{4T\,t\,u}{r^2}$$

(5-13)

where

S   is storage coefficient (L^3/L^3),

t   is elapsed time since pumping began (T),

r   is radial distance from the pumping well to an observation well (L), and

u   is the Theis equation parameter.

Theis Type-Curve Methodology

The Theis type-curve method is a graphical solution to equations 3-1 and 3-2. The type-curve method employs two superimposed curves, $W(u)$ vs. $1/u$ and s vs. t, where s is the drawdown ($h_0 - h$). The first plot is called the Theis type curve and represents the theoretical response of a confined aquifer to a constant pumping stress. The plot of $W(u)$ vs. $1/u$ is a graphical representation of the Theis integral seen in equation 3-1. The second plot is the measured drawdown data from an observation well at a known distance from the pumping well versus the time since the start of the pumping stress.

Both of these relations are plotted on identical scale log-log axes and superimposed on one another (Figure 5.15). The curves are shifted laterally and/or vertically relative to each other into a position where the plot of measured field data directly overlies the theoretical response curve ($W(u)$ vs. $1/u$). These adjustments are made so that the axes of both graphs remain parallel with one another and the data points lie as close as possible to the type curve. Once the theoretical and the field data are in a proper matching position, any arbitrary match point is chosen on the overlapping graphs and values of $W(u)$ and $1/u$ from the theoretical response curve and s and t are recorded from the measured field data plot (Figure 5.15).

The calculations are easier when the match point is selected at intersections of major axes of $W(u)$ and $1/u$ (i.e., where $W(u)$ and $1/u$ are equal to even powers of 10). Thus, it is not necessary to choose a match point on the Theis type curve. Two of the match point values, $W(u)$ and s, are substituted into equation 5-12 and used to compute transmissivity (T). Once a transmissivity value is calculated, it is substituted into equation 5-13 to solve for storage coefficient (S) using the match point values of t and u. Be sure to convert the value of $1/u$ to u.

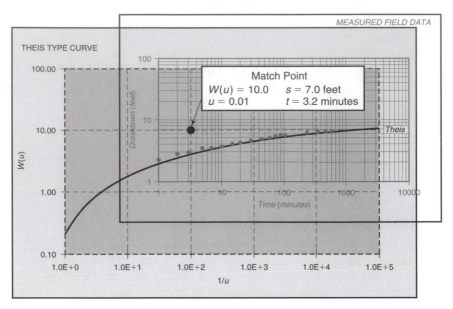

Figure 5.15. Plot overlay with $W(u)$ vs. $1/u$ and drawdown vs. time for Theis type-curve methodology and corresponding match point value.

The Theis type-curve method using a match point as described above is commonly used when the two overlain graphs are in printed form. However, in the *Excel* problem, the axes of the theoretical plot of the Theis type curve are not displayed and no match point is required. Initial values of transmissivity (T) and coefficient of storage (S) are determined by the user employing scroll bars to adjust the position of the Theis type curve (see *Excel Tips*). Thereafter, you make adjustments to the values of T and S that automatically adjust the position of the Theis type curve. Your task in this problem is to make these adjustments until you believe the optimum fit is achieved between the measured field data plot and the theoretical Theis curve (Figure 5.16). Values of T and S are displayed in the worksheet.

Determining the optimal fit of the Theis type curve to the drawdown data takes practice and patience. When fitting the curve, weight the early-time data less than the later time data because the late-time data more closely represent the theoretical drawdown equation on which the type curve is based (Kruseman and de Ridder, 1990). Two of the assumptions on which the Theis equation is based are that water is released instantaneously from storage and that the pumping rate is a constant. However, at the beginning of the test (early-time data) there may be borehole storage effects, especially in large-diameter production wells, and non-instantaneous release of water from aquifer storage. The pumping rate may fluctuate slightly at the beginning of the test because the pump must adjust to the rapid change in head in the pumped well. However, these early-time data are important in maximizing the accuracy of the aquifer parameter determination. The rapid rate of head changes in early time helps define the transmissivity value from the Theis type curve. Figure 5.16 illustrates a Theis type-curve fit to drawdown data. Note that the first data point at 6 seconds does not fit the type curve but that the fit becomes progressively better, so that by 42 seconds the data match the Theis type curve.

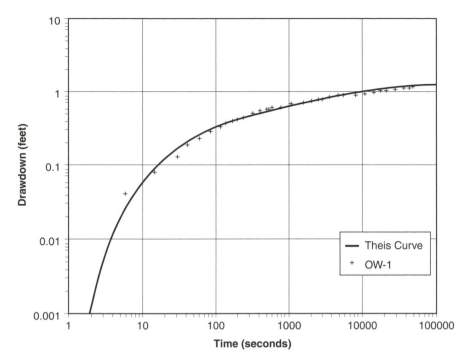

Figure 5.16. Appropriate fit of Theis type curve to water levels measured in an observation well.

Drawdown can be measured in several observation wells during an aquifer test and analyzed by the Theis type curve method to determine aquifer parameters. Each set of observation well data will produce an estimate of T, K, and S when analyzed. Because radially horizontal flow is one of the assumptions of Theis analysis, two observations wells that are orthogonal to one another with respect to the pumping well will yield estimates of T and K in orthogonal directions. Thus, anisotropy, if present, can be observed. However, different estimated parameter values derived from different observation-well data may also reflect the heterogeneity of the aquifer. Therefore, data from many wells can be used to estimate different aspects of the aquifer.

In applying this method for estimating values of aquifer parameters, one must always use consistent units. Thus, the units of time and distance must be the same for all the parameters used. For example, if drawdown is measured in feet, the radial distance to the observation well and the aquifer thickness must be expressed in units of feet and the pumping rate of the well must be expressed in cubic feet. Time units must be similarly consistent.

Excel Tips

- This worksheet contains a macro used to make calculations for the Theis equation. Your security setting within *Excel* should be set to "Medium," which allows you to run associated macros. The security setting can be modified within *Excel* under the Tools >> Macro >> Security menu, which brings up the dialog box shown below.

- If the security setting within *Excel* is set to "Medium," the following dialog box will appear when you try to open the worksheet. To allow full functionality of the worksheet, select "Enable Macros" in this dialog box, as seen below.

- If you receive a message about digital signatures and certificates, or if the macros in the worksheet do not work properly, you may need to change your security settings to "Medium" to allow their operation.
- Exponent symbol: $^\wedge$ raises a number to the designated power (e.g., $3^\wedge 2 = 9$).

- Other intrinsic *Excel* functions are listed and explained by clicking the f_x button.
- Use the scroll bars and arrows to adjust transmissivity and storage coefficient values as seen below.

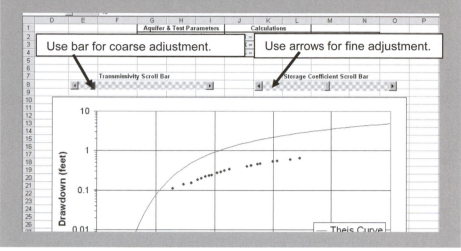

Parameter	Definition
h	Hydraulic head after elapsed time (L)
h_0	Initial hydraulic head (L)
Q	Constant discharge from the pumping well (L^3/T)
r	Radial distance from pumping well to observation well (L)
s	Drawdown or ($h_0 - h$) (L)
S	Storage coefficient (L^3/L^3)
t	Elapsed time since the pumping began (T)
T	Transmissivity of the aquifer (L^2/T)
u	Theis equation parameter
$W(u)$	Theis well function

Table 5.5 Parameter Definition Table – Chapter 5, Problem 4

Problem 5

COOPER-JACOB STRAIGHT-LINE METHOD OF AQUIFER TEST ANALYSIS

Englewood Well Field, Southwest Sarasota County, Florida

Overview

The Cooper-Jacob straight-line method (Cooper and Jacob, 1946) is a simplification of the Theis equation that can be used under certain conditions. With long pumping durations, large transmissivities, and small radial distances to observation wells, the higher power terms in the Theis infinite series (equation 3-3) become negligible (see Chapter 3 – Problem 2). Under certain conditions, this relation enables the infinite series to be truncated such that a simpler solution can be formulated. The criterion used to determine if the infinite series can be truncated is based on small values of the parameter u: if $u < 0.01$, then the Cooper-Jacob method may be used (Kruseman and deRidder, 1990). For a given aquifer test and radial distance to an observation well, the "u criterion" can be expressed in terms of time (t), such that measured values of time and drawdown can be used only when the following equation is satisfied.

$$t \geq \frac{25\,r^2\,S}{T} \quad for \; u < 0.01 \tag{5-14}$$

To determine if the Cooper-Jacob straight-line method is appropriate, one must first calculate values of S and T using the method and then substitute those values into equation 5-14 to see if the "u criterion" is met. Thus, one must use the method to compute possible T and S values before determining if the results meet the "u criterion" to validate use of the method.

There are two types of analyses that can be used with the Cooper-Jacob method. One is time-drawdown analysis where measured values of drawdown versus time are plotted for a single observation well. The other type of analysis is distance-drawdown, where drawdowns measured at a specific time in three or more observation wells are plotted versus the distances of the observation wells from the pumped well. For the time-drawdown method, values of transmissivity and the coefficient of storage are calculated with the following relations.

$$T = \frac{2.3\,Q}{4\,\pi\,\Delta(h_0 - h)} \tag{5-15}$$

$$S = \frac{2.25\,T\,t_0}{r^2} \tag{5-16}$$

where

T	is the transmissivity of the aquifer (L^2/T),
Q	is the constant pumping rate (L^3/T),
$\Delta(h_0 - h)$	is the drawdown per log cycle of time on the time-drawdown plot (L),
S	is the storage coefficient (L^3/L^3),
r	is the radial distance from the pumping well to the observation well (L), and
t_0	is the time at which the straight-line intersects the zero-drawdown axis (T).

Figure 5.17 shows how to determine the value of $\Delta(h_0 - h)$ from the graph of time versus drawdown. The value of $\Delta(h_0 - h)$ is the change in hydraulic head during one log cycle of time using the best-fit line. The value of t_0 is the value of time where the best-fit line intersects the zero-drawdown axis.

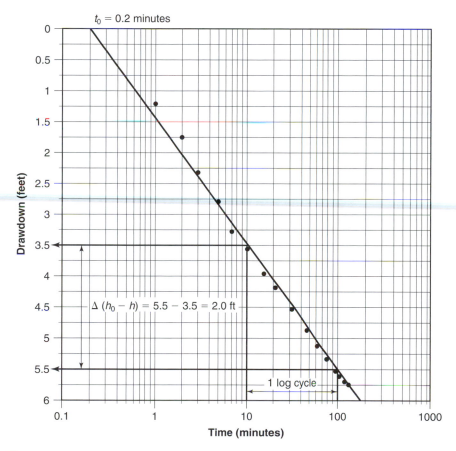

Figure 5.17. Example of time versus drawdown plot for use with the Cooper-Jacob straight-line method.

For the distance-drawdown method, which requires a minimum of three observation wells, a plot of distance to the observations wells from the pumping well versus drawdown in the observation wells at a specific time is made and values of transmissivity and the coefficient of storage are calculated with the following relations.

$$T = \frac{2.3 Q}{2 \pi \Delta(h_0 - h)} \tag{5-17}$$

$$S = \frac{2.25 T t}{r_0^2} \tag{5-18}$$

where

T	is the transmissivity of the aquifer (L^2/T),
Q	is the constant pumping rate (L^3/T),
$\Delta(h_0 - h)$	is the drawdown per log cycle of distance on the drawdown-distance plot (L),
S	is the storage coefficient (L^3/L^3),
r_0	is the distance at which the straight line intercepts the zero-drawdown axis (L), and
t	is the elapsed time since pumping began when the drawdowns were measured (T).

These analysis methods have all the same assumptions as the Theis equation.

Excel Tips

- PI() inserts the constant π into an equation.
- Cell anchor: $column$row anchors the cell (column, row) in a formula. Performing a relative copy will maintain the anchored cell within the formula and change only the other non-anchored cells in the formula (e.g., =$D$12+D2 will anchor cell D12 and not cell D2 during a relative copy).
- Exponent symbol: ^ raises a number to the designated power (e.g., 3^2 = 9).
- Other intrinsic *Excel* functions are listed and explained by clicking the f_x button.

Parameter	Definition
$\Delta(h_0 - h)$	Drawdown per log cycle of time/distance on the drawdown-time/distance plot (L)
r	Radial distance from the pumping well to the observation well (L)
r_0	Distance at which the straight line intercepts the zero-drawdown axis (L)
Q	Constant pumping rate (L^3/T)
s	Drawdown (L)
S	Storage coefficient (L^3/L^3)
t	Time since the pumping began (T)
T	Transmissivity of the aquifer (L^2/T)
t_0	Time where straight line intersects the zero-drawdown axis on the time-distance plot (T)
u	Theis equation parameter

Table 5.6 Parameter Definition Table – Chapter 5, Problem 5

CHAPTER 6

CONTAMINANT TRANSPORT AND REMEDIATION

Principles & Concepts

Many recent advances in the field of groundwater hydrology are associated with understanding the physical and chemical controls of contaminant transport through geologic materials. Better conceptualization and quantitative description of contaminant transport processes enable scientists and engineers to more accurately predict where contamination may exist and evaluate various remediation technologies. Quantitative descriptions of these processes have proven to be useful in addressing groundwater contamination problems.

This chapter presents problems dealing with the physical and chemical transport of solutes through porous materials and examines simple remediation techniques. The objective of the chapter is to examine the physical and chemical controls of contaminant transport including advection, diffusion, dispersion, and chemical retardation. Problems 1 and 2 use analytical solutions to the solute transport equation to address the processes that move contaminants downgradient in a groundwater flow system. Problem 3 examines the design of a contaminant recovery well in a pump-and-treat system for the remediation of groundwater pollution. The worksheet uses equations delineating the capture zone of a pumping well to compute the area contributing to a contaminant extraction well and the sensitivity of capture-zone geometry to aquifer parameters and design parameters.

Transport Processes

Understanding the natural processes that influence the transport of dissolved contaminants in groundwater is important in determining the fate of these contaminants in the subsurface environment. One of these processes is advection, whereby dissolved contaminants are transported in moving groundwater. In many groundwater environments, advection accounts for the majority of contaminant movement. Other physical transport processes include diffusion, whereby dissolved contaminants are transported from areas of high concentration to areas of low concentration. This movement is the result of random molecular motion due to the kinetic energy of the solute. The third physical process of mass transport is dispersion, which is the spreading of a contaminant because of microscopic variations in flow velocities within the porous medium. Thus, contaminants emanating from a continuous source or a pulse source (spill) will migrate down the hydraulic gradient by advection and will also become more dilute downgradient because of diffusion and dispersion.

Figure 6.1 shows the chloride and sodium plumes at Otis Air Force Base on Cape Cod, Massachusetts, migrating down the hydraulic gradient because of advection and becoming more dilute near the edges of the plume owing to dispersion and diffusion.   We can represent these three physical transport processes mathematically in one dimension using the Ogata and Banks (1961) equations presented in Problem 1.

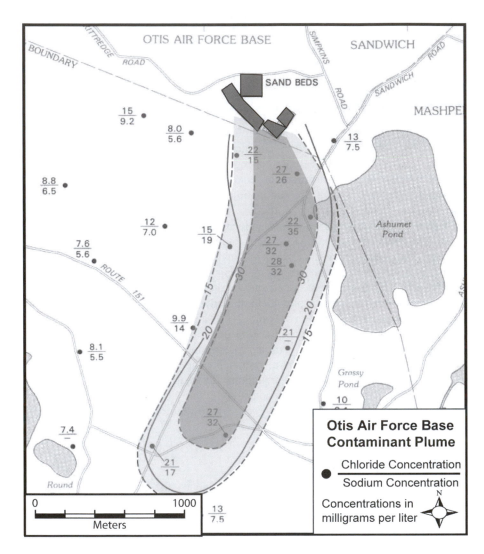

Figure 6.1. Sodium and chloride plumes from sewage disposal sand beds at Otis Air Force Base in eastern Massachusetts (modified from LeBlanc, 1984).

Quantitative Description of Plumes

Concentrations in a contaminant plume can be described by using concentration profiles and breakthrough curves.   Concentration profiles are graphical descriptions of relative concentrations versus distance along a longitudinal transect (Figure 6.2) through the plume.   Figure 6.3 illustrates a concentration profile in which the relative concentration (C/C_0) is the measured concentration at a given distance from the source (C) divided by the initial concentration at the source (C_0). Figure 6.4 shows a breakthrough curve that depicts the change in relative concentration of a contaminant at one location (e.g., a monitoring well) with time as the plume migrates past the measurement location.

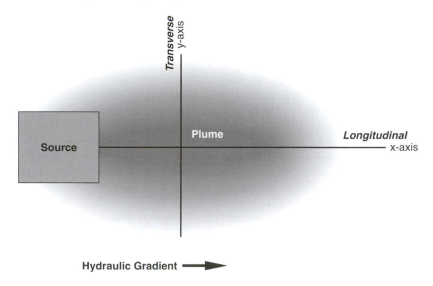

Figure 6.2. Plan view of a contaminant plume moving downgradient to the right from a continuous source and becoming more dilute on the edges of the plume in both the longitudinal and transverse directions.

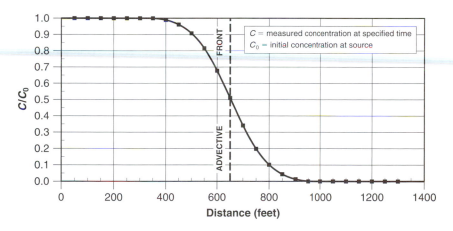

Figure 6.3. A concentration profile is a way to illustrate the change in relative concentration (C/C_0) with distance at a specific time after the release of the contaminant.

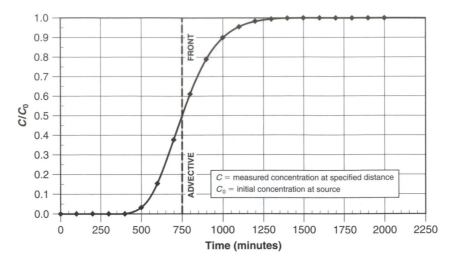

Figure 6.4. A breakthrough curve illustrates the change in relative concentration (C/C_0) with time at a specific distance from the contaminant source.

The advective front of a contaminant plume is the distance that the dissolved contaminant travels under the influence of advection only (i.e., no dispersion or diffusion). This advective front is represented on concentration profiles and breakthrough curves as the point at which C/C_0 is 0.5—the concentration is half the initial concentration. Contaminants ahead of this point are traveling faster than the average linear flow velocity owing to dispersion and diffusion, and contaminants behind this point are traveling more slowly than the average linear flow velocity owing to dispersion and diffusion. This advective front on Figures 6.3 and 6.4 can be seen on the concentration profile and breakthrough curve at C/C_0 at 0.5 (approximately 550 feet in Figure 6.3 concentration profile example and 750 minutes in Figure 6.4 on the breakthrough curve example).

In addition to the physical transport processes described above, chemical transport processes also modify contaminant concentrations. Surface chemical reactions at water/mineral interfaces cause wholesale changes to groundwater chemistry. One of these chemical reactions is sorption, whereby hydrophobic (water-hating) contaminant compounds preferentially partition into solid organic matter. This organic matter, which is commonly present in aquifers as films on individual mineral grains, allows for these nonpolar contaminant compounds to sorb into nonpolar environments. The process of sorption tends to slow the transport of these hydrophobic contaminant compounds. We account for this process mathematically in the transport equation by using a chemical retardation factor. Chemical retardation decreases both the advective velocity and the dispersive characteristics of a contaminant plume, as illustrated in Figure 6.5. In this example, there is less advection of the chemically retarded plume, as indicated by the slowing of the advective front of the two concentration profiles. Dispersion is also limited in the retarded species, as observed by the less inclined concentration profile.

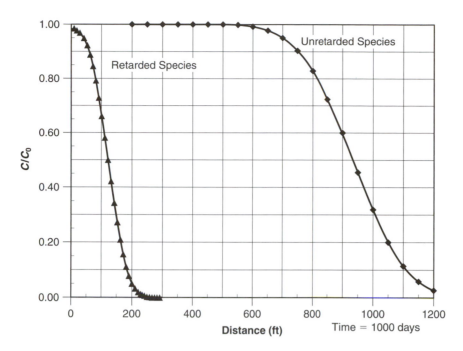

Figure 6.5. Concentration profiles illustrating the change in relative concentration (C/C_0) versus distance for a contaminant that does not chemically react with the sediment (unretarded) and a chemically retarded contaminant that are transported in the same groundwater flow system.

Contaminant Remediation

Remediation of contaminant plumes can take many forms, one of which is removal and treatment of contaminated water (pump and treat). This method requires the design of effective and efficient systems that pump the contaminated water from the aquifer and treat it at the surface. To be effective and efficient, the capture zone of a contaminant recovery well must closely circumscribe the contaminant plume (Figure 6.6). Overestimation of the capture zone incurs the undesired expense (inefficiency) of treating clean water, whereas underestimation of the capture zone enables contaminants to escape downgradient (ineffectiveness). Understanding the sensitivity of the variables that influence the geometry of capture zones is examined in Problem 3, which involves trichloroethylene (TCE) contamination at a well field in Wooster, Ohio. The simple equations used in this problem are based on assumptions that include the homogeneous distribution of hydraulic conductivity, constant aquifer thickness, a uniform hydraulic gradient, and a constant pumping rate of the recovery well.

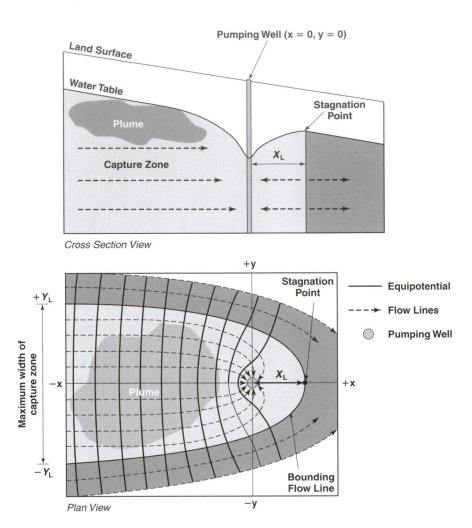

Figure 6.6. Geometry of a pumping well capture zone showing the distance to the stagnation point (X_L) and the maximum width of the capture zone $(-Y_L$ to $+Y_L)$ along with flow lines and equipotential lines.

Contaminated groundwater extracted from an aquifer generally requires on-site treatment or disposal. In this chapter, air-stripping technology is introduced as a method for treating volatile organic chemicals (VOCs). Air stripping takes advantage of the low vapor pressure of many VOCs such that many of these compounds would rather partition from the liquid to gas phase. In an air-stripping tower, contaminated water cascades down through a steel tower filled with small, plastic devices that maximize the surface area of the falling water. At the same time, air is blown up through the column to volatilize ("strip") the VOCs from the water (Figure 6.7). Air-stripping towers are an efficient method for removing VOCs such as TCE and perchloroethylene (PCE) from contaminated groundwater but are not an efficient method for removing contaminants with low vapor pressures.

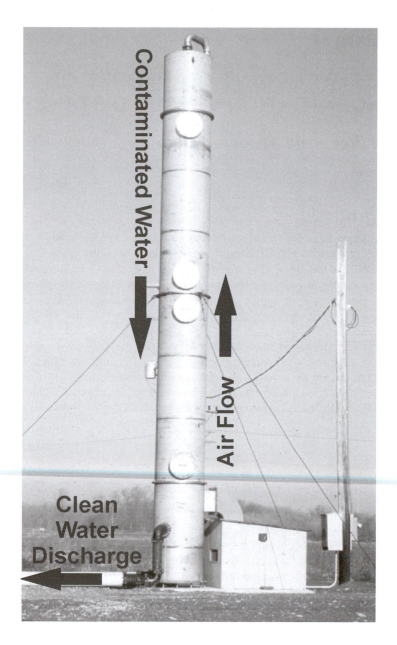

Figure 6.7. Air-stripping tower designed to remove VOCs from contaminated groundwater.

Problem 1

1-D CONTAMINANT TRANSPORT OF TCE (CONCENTRATION VERSUS DISTANCE)

South Well Field, Wooster, Ohio

Background

This exercise uses an analytical solution to a general solute transport equation to determine contaminant concentration versus distance downgradient that the contaminant has traveled. This analytical solution is simplified to address one-dimensional transport, which allows us to more easily examine the physical and chemical processes that control the transport of contaminants in groundwater systems.

It is easy to visualize the importance of the advective groundwater velocity in determining the concentration of a contaminant downgradient of its source. This advective movement is the dominant transport process that moves the contaminant downgradient at the average linear flow velocity. The velocity of groundwater flow can be determined by using the following equation.

$$v_x = \frac{K\ i}{n_e} \tag{6-1}$$

where

v_x is the average linear flow velocity in the x direction (L/T),
K is the hydraulic conductivity of the aquifer (L/T),
i is the hydraulic gradient in the x direction (L/L), and
n_e is the effective porosity or the porosity available for fluid flow (expressed as a decimal).

Understanding the processes that cause a contaminant to disperse and spread in both the longitudinal (x) and transverse (y) directions (Figure 6.2) requires introduction of three additional variables. The coefficient of hydrodynamic dispersion (D_x) describes the overall spreading of a contaminant because of these other physical processes. D_x depends on the velocity of the groundwater flow (v_x) and the dispersivity of the geologic material (α_x), which accounts for the microscopic variation in flow velocities through the pore throats. The molecular diffusion (D^*) is contaminant specific and is also accounted for in the coefficient of hydrodynamic dispersion term. Equation 6-2 describes the relation of these variables.

$$D_x = v_x \ \alpha_x + D^* \tag{6-2}$$

where

D_x is the coefficient of hydrodynamic dispersion in the x direction (L^2/T),

v_x is the average linear flow velocity in the x direction (L/T),

α_x is the dispersivity in the x direction (L), and

D^* is the coefficient of molecular diffusion of the contaminant (L^2/T).

Ogata-Banks Equation

Ogata and Banks, in a U.S. Geological Survey Professional Paper published in 1961, presented a one-dimensional, analytical solution for the solute transport equation. A simplified form of this analytical solution appears below in equation 6-3. By employing the complementary error function (ERFC), the contaminant concentration at any time and any distance downgradient from a constant source can be calculated with the following equation.

argument

$$C = \frac{C_o}{2} \ \mathrm{ERFC} \left[\frac{x - v_x t}{2\sqrt{D_x t}} \right] \tag{6-3}$$

where

C is the concentration of the contaminant (M/L^3),

C_0 is the initial concentration of the contaminant at the source (M/L^3),

x is the distance to the point of interest (L),

v_x is the average linear flow velocity in the x direction (L/T),

D_x is the hydrodynamic dispersion in the x direction (L^2/T), and

t is the time (T).

Scientists and engineers use this equation to simulate the contaminant concentrations illustrated in breakthrough curves and concentration profiles (Figures 6.3 and 6.4). The complementary error function (ERFC) is used in this analytical solution; the argument in brackets is known as β (Beta). Values of ERFC and ERF (error function) are shown in Figure 6.8, for given values of β.

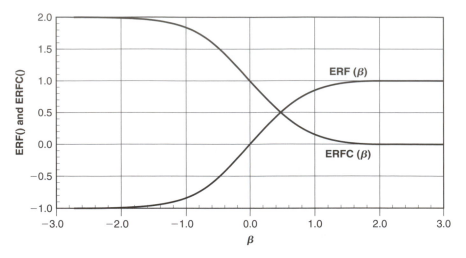

Figure 6.8. Graph of error function and complementary error function behavior from β equal to –3.0 to 3.0.

A useful relation between the error function and the complementary error function is seen below.

$$\text{ERFC } [-\beta] = 1 + \text{ERF } [\beta] \tag{6-4}$$

ERF and ERFC are intrinsic functions in *Excel*. To use these functions within an *Excel* equation, use ERF() for error function and ERFC() for complementary error function. These functions are already programmed into the problem worksheet. However, you will need to program the various terms in the argument of the function (β). See the *Excel Tips* to ensure proper operation of these intrinsic functions. The following two equations provide further information on the ERF and ERFC functions but are not needed to complete the problems.

$$\text{ERF}(\beta) = \frac{2}{\pi} \int_{0}^{\beta} e^{-t^2} \, dt \tag{6-5}$$

$$\text{ERFC}(\beta) = \frac{2}{\pi} \int_{\beta}^{\infty} e^{-t^2} \, dt \tag{6-6}$$

Physical and Chemical Retardation

Chemical retardation of contaminants slows mass transport below the rate of advective transport. Retardation can result from both physical and chemical processes that cause sorption of a contaminant to sediment or mineral surfaces. Part III of this exercise addresses the effect of chemical retardation on the rate and distance that a contaminant plume migrates.

The following equation is the one-dimensional, analytical solution to a transport equation that also accounts for retardation of a chemical species (Ogata and Banks, 1961).

$$C = \frac{C_\circ}{2} \text{ ERFC} \left[\frac{(R_f x - v_x t)}{2 \sqrt{\alpha_x v_x t R_f}} \right] \tag{6-7}$$

where

R_f is the retardation factor and
α_x is the dispersivity in the x direction (L).

The velocity of the retarded contaminant is related to the average linear flow velocity by the following relation. The retardation factor is by definition 1 or greater than 1. Thus, the velocity of the contaminant will be less than the average linear flow velocity, such that

$$v_c = \frac{v_x}{R_f} \tag{6-8}$$

where

v_c is the velocity of the contaminant (L/T),
v_x is the average linear flow velocity in the x direction (L/T), and
R_f is the retardation factor.

Determination of the retardation factor (R_f) is based on properties of the aquifer material and properties the contaminant being examined, such that

$$R_f = 1 + \left(\frac{1 - n_e}{n_e} \right) \rho_b K_d \tag{6-9}$$

where
R_f is the retardation factor,
n_e is the effective porosity,
ρ_b is the dry, bulk density of the aquifer material (M/L^3), and
K_d distribution coefficient (L^3/M).

Porosity and dry bulk density are properties of the aquifer materials that relate to the volume of material present and to the surface area available for the sorption process. The distribution coefficient (K_d) refers to the ability of aquifer materials to sorb a specific chemical contaminant.

In this problem, we are concerned with the sorption of TCE onto the sand and gravel aquifer at one of the municipal well fields at Wooster, Ohio. The degree of sorption of synthetic organic chemicals, such as TCE, can be determined by use of the partition coefficient between organic carbon and water and by the amount of organic carbon present in the aquifer. The K_{oc} for TCE is 152 ml/g and the organic carbon content within the sand and gravel aquifer at Wooster is 1% or 0.01. These partition coefficients have been determined for a wide variety of organic pollutants and can be found in most standard hydrogeology textbooks. The organic carbon content of sediments and rocks is commonly derived from laboratory tests. We can relate the distribution coefficient (K_d) to the partition coefficient and organic carbon content using the following relation.

$$K_d = K_{oc} \; f_{oc} \qquad\qquad (6\text{-}10)$$

where

K_d is the distribution coefficient (L^3/M),

K_{oc} is the partition coefficient between organic carbon and water (L^3/M), and

f_{oc} is the fraction organic carbon of the aquifer.

Excel Tips

- To use the ERF and ERFC intrinsic functions in *Excel*, the "Analysis ToolPak" must be installed as an Add-In. If you receive "#NAME?" under the ERFC column within the worksheet, you may not have this "Analysis ToolPak" installed. To install it, use menu Tools >> Add-Ins then check "Analysis ToolPak" (see below). You may be asked to insert your Microsoft Office CD.

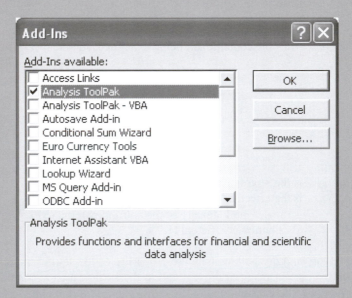

- Cell anchor: $column$row anchors the cell (column, row) in a formula. Performing a relative copy will maintain the anchored cell within the formula and change only the other non-anchored cells in the formula (e.g., =$D$12+D2 will anchor cell D12 and not cell D2 during a relative copy).
- Calculate the average linear flow velocity (v_x) and hydrodynamic dispersion coefficient (D_x) before determining the concentration of the plume at various distances downgradient. For Part III, calculate the distribution coefficient (K_d), retardation factor (R_f), and contaminant velocity (v_c) before calculating the concentrations.

- Error Functions: ERF() and ERFC() are intrinsic functions in *Excel* that return the value of the error function or the complementary error function for the argument inside the parentheses.
- Square Root: SQRT() is an intrinsic function in *Excel* that returns the square root of the quantity inside the parentheses.
- Remember to save your work often. *Excel* has an AutoSave function available under the Tools menu that may be an Add-In function with some versions.

Parameter	Definition
α_x	Dispersivity in the x direction (L)
C	Concentration at distance x after time t (M/V)
C_0	Initial concentration at source (M/L^3)
D_x	Hydrodynamic dispersion coefficient in the x direction
D^*	Coefficient of molecular diffusion of the contaminant (L^2/T)
f_{oc}	Fraction organic carbon in the aquifer material (decimal)
i	Hydraulic gradient (L/L)
K	Hydraulic conductivity of the aquifer (L/T)
K_d	Distribution coefficient (L^3/M)
K_{oc}	Partition coefficient between organic carbon and water (L^3/M)
n_e	Effective porosity (decimal)
R_f	Retardation factor
ρ_b	Dry, bulk density of aquifer material (M/L^3)
t	Time since source became active (T)
v_c	Flow velocity of contaminant (L/T)
v_x	Average linear flow velocity of the groundwater in the x direction (L/T)
x	Distance downgradient from the source (L)

Table 6.1 Parameter Definition Table – Chapter 6, Problem 1

Problem 2

1-D CONTAMINANT TRANSPORT OF TCE (CONCENTRATION VERSUS TIME)

Anne Anderson et al. vs. W.R. Grace et al., Woburn, Massachusetts

Background

This problem examines the concentration versus time relation of a contaminant plume in Woburn, Massachusetts. Using the same analytical solution presented in Problem 1 of this chapter (equation 6-3), you will develop breakthrough curves of relative concentration versus time (instead of distance, as used in Problem 1). Thus, the same parameters are used along with the quantitative relations that examine the relative concentration distribution. For this problem, three locations are chosen downgradient of the source area (800, 1000, and 1200 feet) and the relative concentrations are determined for times from 1 to 2000 days. These calculations produce three concentration breakthrough curves when displayed in graphical form for the three locations. Use equation 6-8 in Part III of the problem to account for chemical retardation along with the associated relations to create these breakthrough curves for the Woburn case study.

Excel Tips

- To use ERF and ERFC intrinsic functions, "Analysis ToolPak" must be installed as an Add-In of *Excel*. If you receive "#NAME?" under the ERFC column within the worksheet, you may not have this "Analysis ToolPak" installed. To install it, use menu Tools >> Add-Ins then check "Analysis ToolPak" (see below). You may be asked to insert your Microsoft Office CD.

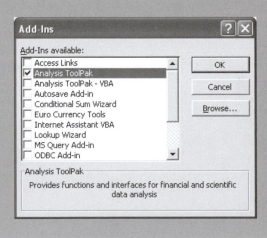

- Cell anchor: $column$row anchors the cell (column, row) in a formula. Performing a relative copy will maintain the anchored cell within the formula and change only the other non-anchored cells in the formula (e.g., =$D$12+D2 will anchor cell D12 and not cell D2 during a relative copy).
- Calculate the average linear flow velocity (v_x) and hydrodynamic dispersion coefficient (D_x) before determining the concentration of the plume at various distances downgradient. For Part III, calculate the distribution coefficient (K_d), retardation factor (R_f), and contaminant velocity (v_c) before calculating the concentrations.
- Error Functions: ERF() and ERFC() are intrinsic functions in *Excel* that return the value of the error function or the complementary error function for the argument inside the parentheses.
- Square Root: SQRT() is an intrinsic function in *Excel* that returns the square root of the quantity inside the parentheses.
- Remember to save your work often. *Excel* has an AutoSave function available under the Tools menu that may be an Add-In function with some versions.

Parameter	Definition
α_x	Dispersivity in the x direction (L)
C	Concentration at distance x after time t (M/L^3)
C_0	Initial concentration at source (M/L^3)
D_x	Hydrodynamic dispersion coefficient in the x direction (L^2/T)
D^*	Coefficient of molecular diffusion of the contaminant (L^2/T)
f_{oc}	Fraction organic carbon in the aquifer material (expressed as a decimal)
i	Hydraulic gradient (L/L)
K	Hydraulic conductivity of the aquifer (L/T)
K_d	Distribution coefficient (L^3/M)
K_{oc}	Partition coefficient between organic carbon and water (L^3/M)
n_e	Effective porosity (expressed as a decimal)
R_f	Retardation factor
ρ_b	Dry, bulk density of aquifer material (M/L^3)
t	Time since source became active (T)
v_c	Flow velocity of contaminant (L/T)
v_x	Average linear flow velocity of the groundwater in the x direction (L/T)
x	Distance downgradient from the source (L)

Table 6.2 Parameter Definition Table – Chapter 6, Problem 2

Problem 3

CAPTURE CURVE ANALYSIS: TCE PLUME

South Well Field, Wooster, Ohio

Overview

Understanding what controls the shape of the capture zone of a pumping well is important not only in water resource management (e.g., wellhead protection issues) but also in the design of effective and efficient contaminant recovery wells. We will explore in this exercise the influence of various parameters on capture-zone morphology, well placement, and operation of a contaminant recovery well. The case study chosen for analysis is the TCE plume at Wooster, Ohio, which also was used as an example in Problem 1 of this chapter. Your task is to use the capture-curve equations to design a TCE recovery well for a portion of the contaminant plume. You will use the confined aquifer equations for the determination of the outer edge of the capture zone, the distance to the stagnation point, and the maximum capture zone width.

Capture Zone Curves in Confined Aquifers

Analytical equations have been derived that delineate the edge of the capture zone of a pumping well for both confined and unconfined aquifers under steady-state conditions (Grubb, 1993). These equations assume homogeneous, isotropic conditions, a uniform regional hydraulic gradient, a constant saturated thickness for confined aquifers, and a constant pumping rate of the well.   The equation used in this exercise describes the outer edge (bounding flow line) of the capture zone of a well in a confined aquifer as

$$x = \frac{-y}{\tan\left[\dfrac{2\pi K b i y}{Q}\right]} \qquad (6\text{-}11)$$

where

x	is the distance parallel to the regional hydraulic gradient (L),
y	is the distance perpendicular to the regional hydraulic gradient (L),
K	is the hydraulic conductivity of the aquifer (L/T),
b	is the saturated thickness of the aquifer (L),
i	is the hydraulic gradient without pumping (L/L), and
Q	is the constant pumping rate of the well (L^3/T).

We can express two solutions to equation 6-11 that describe specific metrics of the capture zone. One of these is the distance to the downgradient stagnation point (X_L), which is defined as the distance from the pumping well

($x = 0$, $y = 0$) to the downgradient edge of the capture zone (Figure 6.9). This distance is expressed by the following equation

$$X_L = \frac{Q}{2\pi Kbi} \tag{6-12}$$

where X_L is the distance from the pumping well at ($x = 0$, $y = 0$) to the downgradient stagnation point.

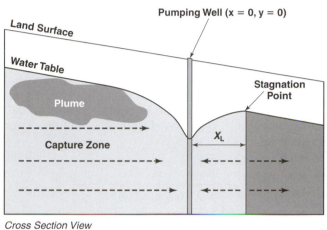

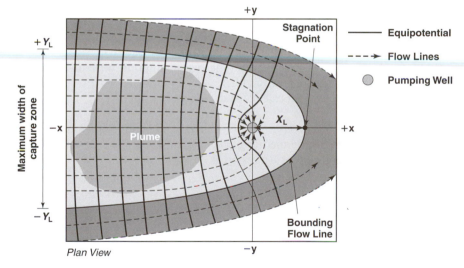

Figure 6.9. Description of capture-zone shape, including distance to the stagnation point (X_L) and the maximum width of the capture zone ($-Y_L$ to $+Y_L$).

The second metric describes the maximum width of the capture zone as x approaches negative infinity (Figure 6.9). This equation is expressed as the half-width of the capture zone.

$$\pm Y_L = \frac{Q}{2Kbi} \tag{6-13}$$

where Y_L is the half width of the maximum capture zone. Thus, the width of the entire capture zone is $2Y_L$.

Capture Zone Curves in Unconfined Aquifers

The analytical equations for capture zone shape in unconfined aquifers are similar to those for confined aquifers; however, in unconfined aquifers the saturated thickness is not constant and the hydraulic gradient is given by two values of hydraulic head, one upgradient and one downgradient of the pumping well. The equation for the outer edge (bounding flow line) of the capture zone in an unconfined aquifer is

$$x = \frac{-y}{\tan\left[\dfrac{\pi K\ (h_1^2 - h_2^2)\ y}{QL}\right]} \tag{6-14}$$

where

h_1 is the upgradient value of hydraulic head (L),
h_2 is the downgradient value of hydraulic head (L),
L is the distance between the two monitoring wells in which h_1 and h_2 were measured (L), and
$\tan()$ is the tangent calculated in radians.

Also similar to the confined aquifer case, two bounding solutions can be expressed to describe specific metrics of the capture zone. The distance to the downgradient stagnation point is expressed as

$$\cdot X_L = \frac{QL}{\pi K(h_1^2 - h_2^2)} \tag{6-15}$$

and the maximum width of the capture zone as x approaches negative infinity is expressed as

$$\pm Y_L = \frac{QL}{K\,(h_1^2 - h_2^2)} \tag{6-16}$$

where Y_L is the half width of the capture zone.

Equations 6-11 and 6-14 at first glance look difficult to solve for y because y appears independently in the numerator and also as one of the parameters within the argument of the tangent function in the denominator. To solve this equation, we employ a Cartesian coordinate system and place the pumping well at the origin ($x = 0$, $y = 0$). If we assume that the direction of the hydraulic gradient is parallel to the *x*-axis from –*x* to +*x*, as shown on Figure 6.9, then a horizontal line through the origin at $y = 0$ is a line of symmetry through the capture zone. This symmetry means that points delineating the bounding flow line will occur as pairs, one having a +*y* value, the other having a –*y* value, for a single value of *x*.

To calculate points describing the bounding flow line using equation 6-11 or equation 6-14, we can select values of *y* and solve for *x* because values of the parameters *K*, *b*, *i*, and *Q* are known from design considerations or field work at the site. The selected values of *y* used to calculate values of *x* need only to range between $Y_L > y > 0$ because of the mirror symmetry.

Available Drawdown and Well Loss

The amount of available drawdown in the pumping well (amount of hydraulic head above the pump hanging in the well) is an important design criterion in determining the pumping rate of a contaminant recovery well or a water supply well. To address this issue more fully, the well efficiency or well loss should be determined. Water flowing across the filter pack and well screen is generally turbulent flow and thus violates the laminar flow assumptions of many of the well hydraulics equations such as the Theis equation. This turbulent flow within the pumping well creates inefficiencies and thus the actual drawdown in the pumping well is greater than the predicted drawdown from equations such as the Theis equation. This increased amount of drawdown because of inefficiencies of the well is termed well loss (i.e., loss of hydraulic head owing to the inefficiencies). Because no well is perfectly (100%) efficient, this inefficiency must be accounted for in the determination of available drawdown in the pumping well itself. The equation used to account for these well losses and determine the drawdown in a pumping well is known as Jacob's equation and is given by

$$s_w \ = \ BQ \ + \ CQ^2 \qquad\qquad (6\text{-}17)$$

where

s_w	is the drawdown in the pumping well (L),
B	is the linear aquifer-loss and well-loss coefficient (T/L^2),
Q	is the constant pumping rate of the well (L^3/T), and
C	is the non-linear well-loss coefficient (T^2/L^5).

Values of the well coefficients *B* and *C* are determined from a step-drawdown test, which is a type of aquifer test in which drawdowns are measured in the pumping well itself while the pumping rate is increased four or five times in a stepwise manner. The analysis of a step-drawdown test to compute values of *B* and *C* for a specific well can be found in Kruseman and de Ridder (1990) and Todd (1980).

Excel Tips

- The intrinsic function for calculation of the π constant is PI().
- The intrinsic function for the calculation of the tangent of an angle is TAN(θ) where θ is the angle in radians.
- Negative y values are the mirror image of the positive y values and are calculated in the worksheet automatically.
- A sign change can be assigned to cells by placing a "−" sign in front of the cell address. For example, −L3 will change the sign of the value in cell L3.
- Large pumping rates (> 500 gpm) will cause the capture curve to be plotted off the chart area.
- Y-positions are based on the X-position range anticipated within exercise.
- Adjust the well location at the bottom of the worksheet to move the plotted well position.
- Remember to save your work often. *Excel* has an AutoSave function available under the Tools menu that may be an Add-In function with some versions.

Parameter	Definition
B	Linear aquifer loss and well-loss coefficient (T/L^2)
b	Saturated thickness of the aquifer (L)
C	Non-linear well-loss coefficient (T^2/L^5)
h_1	Upgradient value of hydraulic head (L)
h_2	Downgradient value of hydraulic head (L)
i	Hydraulic gradient without pumping (L/L)
K	Hydraulic conductivity of the aquifer (L/T)
L	Distance between the two monitoring wells in which h_1 and h_2 were measured (L)
Q	Constant pumping rate of the well (L^3/T)
s_w	Drawdown in the pumping well (L)
x	Distance parallel to regional hydraulic gradient (L)
X_L	Distance from pumping well to downgradient stagnation point (L)
y	Distance perpendicular to regional hydraulic gradient (L)
Y_L	Half width of maximum capture zone (L)

Table 6.3 Parameter Definition Table – Chapter 6, Problem 3

CHAPTER 7
GROUNDWATER CHEMISTRY

Principles & Concepts

Water is the universal solvent and, when allowed to remain in contact with sediments and rocks for extended periods of time, it can create complex geochemical systems. Many of the dissolved constituents that are naturally found in groundwater are the result of dissolution-precipitation, oxidation-reduction, ion exchange, acid-base, and surface reactions. These reactions can considerably change the concentration of dissolved constituents present in a groundwater flow system. This chapter presents some basic tools of aqueous geochemistry that are used by groundwater hydrologists to identify and understand local water resources and regional flow patterns.

Figure 7.1. Fracture-dominated flow through carbonate rock discharges from horizontal fractures and bedding planes along the quarry wall. The light-colored staining of the quarry wall below the bedding plane is from oxidation of the dissolved constituents carried in the groundwater flow system.

The exercises in this chapter focus on low-temperature aqueous geochemistry within the saturated zone. Many water, rock, and sediment interactions control the fate and transport of dissolved constituents in groundwater. We will examine relatively straightforward geochemical systems to better understand how geochemical data are reported and used as analytical tools to interpret leachate migration, geochemical evolution, and the mixing of surface water and groundwater. For more information on groundwater sampling techniques and analysis protocols, see Deutsch (1997), Sanders (1998), Fetter (2001), or Weight and Sonderegger (2001).

Scientists and engineers use a variety of techniques in the field and laboratory to measure water quality and concentrations of dissolved constituents in groundwater systems. The first exercise in this chapter uses the measurement of major ion concentrations to examine the effect of human activities on a local aquifer. Land-use practices can promote the formation of leachate that alters the chemistry of shallow groundwater. As precipitation percolates through the unsaturated zone (and through materials that humans have applied to the land surface), it reacts with soluble compounds. The result is infiltrating water that has some chemical characteristics derived from the material through which it percolated. This chemically altered water, leachate, can reduce groundwater quality in an area. For example, leachate from cattle feedlots usually contains elevated nitrate concentrations from animal wastes; landfills can produce leachate with elevated concentrations of inorganic compounds and total dissolved solids; and leachate from coal or metal mining operations can create high concentrations of heavy metals and have acidic characteristics (acidic mine drainage).

Figure 7.2. View to the west shows the Sawatch Mountain Range in the background and Leadville, Colorado, in the foreground. Throughout the late 1800s this location was the site of gold, silver, and lead mining, which left large waste piles scattered throughout the area. As precipitation percolates through the waste piles, runoff and leachate form that change the chemistry of the local groundwater.

The chemical evolution of groundwater can be observed by comparing the geochemical characteristics of water in recharge areas and in discharge areas in a regional groundwater flow system. In regional flow systems, such as those examined in Chapter 2, groundwater can be in contact with aquifer and confining materials for time periods from years to millennia. These long time periods allow some geochemical reactions, such as dissolution and precipitation, to occur in regional groundwater flow systems to an extent that may not be possible in smaller groundwater flow systems where reaction times are limited by the flow system. Understanding carbonate dissolution and precipitation reactions in their geologic context is of particular interest to the case study presented in Problem 2 of this chapter. In this study, a regional carbonate aquifer system is examined in terms of mineral solubility and thermodynamic equilibrium to understand the geochemical evolution of groundwater within the regional flow system from its recharge area to its discharge locations.

Figure 7.3. Calcite ($CaCO_3$) precipitation forms stalactites and stalagmites, which are a physical manifestation of carbonate chemical equilibrium.

Scientists can use data concerning the chemical composition of groundwater and various graphical techniques, such as Piper diagrams, to identify different types of groundwater. The different types can be classified using the concept of prevalent chemical character, which can tell us about the types of water-rock interactions occurring at different parts of the flow system. Other graphical geochemical tools presented in the second problem can be used to understand the interaction or mixing of water from two distinct sources. In the third exercise, surface water and groundwater mixing is examined using these geochemical tools to determine the amount of induced infiltration between an aquifer and a river. As described in the principles section of Chapter 4, surface water from rivers or lakes can be induced to infiltrate into a shallow aquifer under a pumping stress. By measuring the major cation and anion concentrations of the surface water and comparing those concentrations to groundwater concentrations, scientists can decipher how much of the water discharging from a nearby pumping well is from a surface water source and how much is from a groundwater source. This information is useful to those who protect water resources, especially because surface waters can more easily become contaminated through accidental spills or mismanagement.

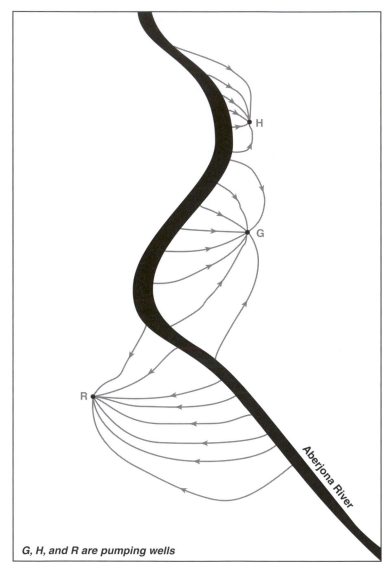

G, H, and R are pumping wells

Figure 7.4. Simulated groundwater flow lines showing induced infiltration from the Aberjona River to the shallow aquifer at Woburn, Massachusetts, created by pumping of wells G, H, and R (modified from Metheny, 1998).

Problem 1

BALANCING MAJOR CATIONS AND ANIONS

Coal Pile Storage, Deepwater Station, New Jersey

Overview

Extended water-rock interaction allows groundwater to establish chemical equilibrium with surrounding mineral phases and to take on the chemical signature of the rocks and sediments through which it passes. This chemical signature usually can be seen in the concentration of dissolved ionic species (cations and anions). There are thousands of possible ionic species in natural waters. However, in most natural groundwaters there are commonly fewer than ten major constituents with concentrations greater than 1 mg/L (Table 7.1).

Major Constituents (> 5.0 mg/L)	Minor Constituents (0.01 – 10.0 mg/L)
Ca^{2+}, Mg^{2+}, Na^+, $H_2CO_3^0$/HCO_3^-, SO_4^{2-}, Cl^-, $H_4SiO_4^0$	B^{3+}, CO_3^{2-}, F^-, $Fe^{2+,3+}$, Sr^{2+}, NO_3^-, K^+

Table 7.1. Major and minor constituents in natural water (modified from Davis and DeWiest, 1966).

Dissolved constituents in groundwater can be described by the relative concentrations of solutes (ions) in the solvent (water). The units are most commonly expressed in either milligrams of solute per liter of solution (mg/L) or milligrams of solute per kilogram of solution (parts per million, ppm). Milligrams per liter is a mass/volume concentration, whereas parts per million is a mass/mass concentration. These two units of concentration are interchangeable for dilute solutions where the total concentration of ions is less than 7,000 mg/L. In freshwater the total concentration of ions is less than 1,000 mg/L. In contrast, average seawater contains 35,700 mg/L of dissolved solids, of which 19,000 mg/L come from Cl^- and 10,500 mg/L from Na^+. For these more concentrated solutions (> 7,000 mg/L), a density correction must be employed to account for the increasing density of the solution as the total concentration of ions increases (see equation 7-1).

$$C_{mg/L} = (\rho_w - TDS) C_{ppm} \qquad (7\text{-}1)$$

where

$C_{mg/L}$	is the concentration in milligrams per liter,
ρ_w	is the density of the water sample in grams per cubic centimeter,
TDS	is the total dissolved solids in grams per cubic centimeter (see discussion below), and
C_{ppm}	is the concentration in parts per million.

Some trace constituents in groundwater occur in smaller concentrations than are easily expressed in mg/L or ppm. These concentrations are generally expressed in micrograms per liter (μg/L) or parts per billion (ppb).

$$1\,ppm = 1,000\,ppb$$
$$1\,mg/L = 1,000\,\mu g/L$$

Two other units used to describe concentrations of dissolved ions in groundwater are molarity (M) and molality (m), because chemical reactions and thermodynamic data such as equilibrium constants (K) are expressed in terms of molarity and molality. For example, the dissolution of calcite is expressed in molarity where one mole of $CaCO_{3(S)}$ dissolves into one mole of Ca^{2+} and one mole of CO_3^{2-}, such that the following relation is satisfied.

$$CaCO_{3(s)} \Leftrightarrow Ca^{2+} + CO_3^{2-} \quad K = 10^{-8.4}$$

Molarity is defined as the moles of solute in one liter of solution, whereas the molality of a solution is defined as the moles of solute in one kilogram of solvent. As is true for the previously discussed concentration units for dilute solutions, these units are interchangeable. Molarity is calculated using the following equation.

$$C_M = \frac{C_{mg/L} \times 0.001}{FW} \tag{7-2}$$

where

C_M	is the molar concentration,
$C_{mg/L}$	is the milligrams per liter concentration, and
FW	is the formula weight or atomic weight.

Cation – Anion Balance

Most groundwater samples are electrically neutral with respect to the concentrations of negatively charged ions (anions) and positively charged ions (cations) in solution. The balance between anions and cations provides a way to check the accuracy of chemical analyses. If the quantity of cations does not equal the quantity of anions, then some ion quantities are missing or some experimental error is present. To determine if a water sample is balanced, we must express the ion concentrations in milliequivalents per liter to account for differences in valences and formula weights. This concentration is expressed by the following formula.

$$C_{meq/L} = C_{mg/L} \left(\frac{V}{FW}\right) \tag{7-3}$$

where

$C_{meq/L}$	is the concentration in milliequivalents per liter,
$C_{mg/L}$	is the concentration in milligrams per liter,
FW	is the formula weight, and
V	is the valence of the ion.

Milliequivalents per liter (meq/L) is equivalent to milliequivalents per kilogram (meq/kg) for dilute concentrations. Summation of the concentrations of cations in milliequivalents per liter should balance the concentrations of anions in milliequivalents per liter. The degree to which this balance is achieved can be determined as the percent difference of the total cation and total anion concentrations (see equation 7-4).

$$\text{Percent Difference} = \frac{\sum \text{cations} - \sum \text{anions}}{\sum \text{cations} + \sum \text{anions}} \times 100 \qquad (7\text{-}4)$$

If the result is positive, then excess cations or insufficient anions exist in the water analysis. If the result is negative, then insufficient cations or excess anions exist in the analysis. An acceptable error in terms of the percent difference is usually defined as less than 5% but preferably it should be less than 2% (Hem, 1985). Reasons for an imbalance can include any of the following (Deutsch, 1997):

1. missing analysis of major dissolved species,
2. laboratory error (or inaccurate analysis),
3. sample contains particulates that dissolve into solution during analysis,
4. minerals precipitate out of solution during analysis, and
5. ionic species not used in the balance calculation.

Filtering of samples in the field is important to remove particulates (suspended solids) before samples are treated with acid or with other preservation techniques. Filtering is usually achieved by using a 0.45-micrometer pore size filter to remove particulates such as metal oxides, clays, and calcite particles. Solid particulates that remain in the sample during acid treatment and other manipulations can introduce errors into ion concentrations. For example, particulate calcite in the sample during titration analysis will result in excess HCO_3^- and CO_3^{2-} being reported for the sample.

Other Water Quality Parameters

Total Dissolved Solids

The total dissolved solids (TDS) content of a water sample can be analytically determined by evaporating a filtered (passed through a 0.45-μm filter) water sample and measuring the residual solids. The computed TDS is the summation of the concentrations in milligrams per liter of all the cations and anions. Major analytical or computation error can be detected by comparing the analytical TDS value with the computed TDS value. To increase the accuracy of this calculation, bicarbonate ions (HCO_3^-) are converted to carbonate in the solid phase. In this process, about half of the bicarbonate is assumed to be volatilized as H_2O and CO_2. Thus, in the calculated TDS, the bicarbonate concentration would be converted to carbonate. This conversion is accomplished by multiplying the measured bicarbonate concentration by 0.4917 to determine how much carbonate would exist from the bicarbonate source. The calculated TDS should be within 10–20 mg/L of the laboratory TDS (Hem, 1985).

Hardness

The concept of hardness as a water quality parameter has been used for thousands of years. Today the term hardness refers to the ability of water to form scales (encrustations) when it is heated or to prevent the sudsing of soap. Hardness is due to the presence of calcium and magnesium cations, which form insoluble compounds with soap and thus prevent sudsing. These insoluble compounds also form in water pipes and equipment when water high in these cations is heated. Hardness is calculated relative to calcite ($CaCO_3$) by the following equation.

$$C_H = C_{meq/L} \ x \ \frac{FW_{calcite}}{V_{carbonate}}$$

(7-5)

where

C_H	is the concentration of hardness as $CaCO_3$ in mg/L,
$C_{meq/L}$	is the concentration of Ca^{2+} and Mg^{2+} summed in meq/L,
$FW_{calcite}$	is the formula weight of calcite (100 g/mole), and
$V_{carbonate}$	is the valence of carbonate (–2 charge).

Hardness Range (mg/L of CaCO₃)	Description
0–60	Soft
61–120	Moderately hard
121–180	Hard
> 180	Very hard

For domestic purposes, hardness above 100 mg/L begins to cause usage problems. Hardness values can become much higher in areas where the water is in extended contact with limestone or gypsum-rich rocks ($CaSO_4 \cdot 2H_2O$). In these cases, hardness values can reach 200–300 mg/L.

pH

The pH of a water sample is a measure of the free, uncomplexed hydrogen ion concentration and is expressed by the following equation.

$$pH = -\log_{10}[H^+]$$

(7-6)

where $[H^+]$ is the activity of hydrogen ion (moles/kg). The pH of water is considered to be a master variable because hydrogen ions participate in most chemical reactions affecting water composition, such as dissolution-precipitation reactions and adsorption-desoprtion reactions. A pH of 7.0 is considered neutral. Values greater than 7.0 are basic and values less than 7.0 are acidic.

The buffering capacity of water, which is measured by adding an acid or base to a sample, is a parameter known as alkalinity (acid-neutralizing capacity) and acidity (base-neutralizing capacity). Alkalinity measures all species in water that can potentially neutralize hydrogen ions. These species commonly include CO_2, HCO_3^-, CO_3^{2-} but can also include $H_3SiO_4^-$, OH^-, and organic ligands (Hem, 1985). Alkalinity is reported in units of mg/L as $CaCO_3$ and is used to determine the carbonate ion concentration (Deutsch, 1997). The acidity of a water sample is not a commonly measured parameter because it cannot be used to calculate the carbonate ion concentration, as is the case for alkalinity.

Specific Conductance

Conductance or specific conductance (SpC) is a parameter usually measured in the field when a water sample is collected. It is defined as the ability of a water sample to conduct an electrical current. In general, the more dissolved ions present in the water, the higher the specific conductance. Because specific conductance is easy to measure, it is used as a proxy for the TDS of the water. However, specific conductance depends on the types of dissolved ions present in the sample, so it represents an approximation of the TDS. Electrical resistance is measured in ohms, and specific conductance, which is the reciprocal, is measured in mhos (ohm spelled backward). In Système International (SI) units, specific conductance is measured in micromhos per centimeter (μmho/cm) at 25°C or microsiemens (μS) [1 μmhos/cm = 1 μS]. It is related to TDS through the following general relation.

$$TDS = SpC \, x \, A \qquad (7\text{-}7)$$

where SpC is the specific conductance in microsiemens and A is a coefficient that ranges from 0.55 to 0.75, depending on the types of ions in solution (Hem, 1985).

Acidic Mine Drainage

Metal mining and coal mining commonly produce large volumes of inorganic contaminants because pyrite (FeS_2) is a widespread gangue mineral left in tailing piles and in storage piles and exposed in the mine. Oxidizing conditions allow higher mobility and concentrations of constituents that pose a hazard to ecosystems and human health. Pyrite and other sulfide minerals form under reducing (anoxic) conditions in the subsurface. When these materials are brought to the surface through coal and metal mining activities, the pyrite becomes unstable and oxidizes. The general oxidation reaction is as follows.

$$4FeS_2 + 15O_2 + 14H_2O \rightarrow 4Fe(OH)_{3(s)} + 8SO_4^{2-} + 16H^+ \qquad (7\text{-}8)$$

Pyrite   *Ferrihydrite*

This oxidation reaction converts ferrous iron (Fe^{2+}) in the pyrite to ferric iron (Fe^{3+}) in a ferrihyrite mineral known as "yellow boy" because of the distinctive yellow color it produces in some acidic mine drainage. The reaction also produces excess sulfate ions and hydrogen ions. Note that for every 4 moles of pyrite consumed, 8 moles of sulfate and 16 moles of hydrogen are produced. As a result, runoff from coal and tailings piles can contain high concentrations of sulfuric acid (H_2SO_4). This water may lower the pH of surrounding water, depending on the buffering capacity and natural acidity of the groundwater system. The acidic runoff can be neutralized by calcite and other carbonate minerals within the system. Equation 7-9 illustrates the neutralization reaction of acid (H^+) and the resulting increase in concentrations of Ca^{2+} and HCO_3^-.

$$H^+ + CaCO_3 \rightarrow Ca^{2+} + HCO_3^- \qquad (7\text{-}9)$$

Oxidation of pyrite in the unsaturated zone or on the Earth's surface is catalyzed by the iron-oxidizing bacterium *Thiobacillus ferrooxidans*. This bacterium promotes oxidation reactions where oxygen is abundant. Therefore, this catalysis is absent in the saturated zone below the water table. The lower pH along with the bacterial activity both prompt Fe^{3+} in solution to form acidic mine drainage (Deutsch, 1997).

Excel Tips

- SUM() sums the cells designated within the parentheses.
- Cell anchor: $column$row anchors the cell (column, row) in a formula. Performing a relative copy will maintain the anchored cell within the formula and change only the other non-anchored cells in the formula (e.g., =$D$12+D2 will anchor cell D12 and not cell D2 during a relative copy).
- Remember to save your work often. *Excel* has an AutoSave function available under the Tools menu that may be an Add-In function with some versions.

Parameter	*Definition*
$C_{mg/L}$	Solute concentration in milligrams per liter (M/V)
C_{ppm}	Solute concentration in parts per million (M/M)
C_M	Solute concentration in molarity (mole/V)
$C_{meq/L}$	Solute concentration in milliequivalents per liter (milliequivalents/V)
C_H	Concentration of hardness as $CaCO_3$ (M/V)
FW	Formula weight (M/mole)
TDS	Total dissolved solids (M/V)
V	Valence charge of an ion
ρ_w	Density of the water (M/V)

Table 7.3 Parameter Definition Table – Problem 1

Problem 2

CHEMICAL EVOLUTION OF GROUNDWATER

Silurian-Devonian Carbonate Aquifer in Central Ohio

Overview

Water-rock chemical interactions and mineral alterations in natural settings are different forms of chemical weathering. One important reaction explored in this exercise is the dissolution and precipitation of minerals within an aquifer. In particular, we examine some simple equilibrium reactions that occur in a regional groundwater flow system where the water is in contact with the aquifer matrix for an extended time. We examine equilibrium and disequilibrium solutions but not the kinetics of the reactions involved.

Geochemical evolution of waters within large sedimentary basins is documented by several studies (Back and Hanshaw, 1965; Plummer, 1977; Freeze and Cherry, 1979; Hem, 1985). Waters generally increase in TDS along regional flow paths as groundwater moves from recharge to discharge locations. Freeze and Cherry (1979) discuss a benchmark paper that considered more than 10,000 chemical analyses in Australia (Chebotarev, 1955); the author concluded that along a regional flow path, groundwater evolves toward the composition of seawater. The dominant anion concentrations take the following stages.

$$HCO_3^- \to \left(HCO_3^- + SO_4^{2-}\right) \to \left(SO_4^{2-} + HCO_3^-\right) \to \left(SO_4^{2-} + Cl^-\right) \to \left(Cl^- + SO_4^{2-}\right) \to Cl^-$$

Bicarbonate (HCO_3^-) is the dominant anion in recharge areas in part because of aggressive dissolution of calcite and dolomite by CO_2-rich rainwater. This increased CO_2 concentration is the result of decomposing organic matter that provides an opportunity for lower pH soil water to be created. As additional gypsum deposits are encountered along the flow paths, the higher solubility allows increased sulfate (SO_4^{2-}) concentrations to develop. The evolution to chloride (Cl^-)-dominant water results from contact with evaporate deposits of halite or sylvite in many deep sedimentary basins. These minerals have solubilities several orders of magnitude greater than calcite, dolomite, gypsum, and anhydrite. Thus, they easily dissolve. The chemical evolution of groundwater in large sedimentary basins is unique to each basin because of the structural evolution of the basin or the mineralogic and hydrologic characteristics of the basin (Freeze and Cherry, 1979). Plummer (1977) discusses the need for hydrologic, mineralogic, and water chemistry data to define reaction models in the chemical evolution of natural waters. Multiple mechanisms or reaction models may exist within these complex flow systems to explain the observed mass transfer along regional flow paths. More complex analyses, including the use of sulfur and other isotopes, are usually necessary to understand geochemical evolution in regional aquifers such as the Floridan and Edwards aquifers in Florida and Texas, respectively (Plummer, 1977; Rye et al., 1981).

Expression of Concentration

Concepts relating to the conversion of concentration values from milligrams per liter to milliequivalents per liter (meq/L) are discussed in Problem 1 of this chapter in the *Reference Book*. Conversion factors for changing concentration from mg/L to meq/L are used in the *Excel* spreadsheet. It follows from equation 7-3 that the conversion factor has the following relation.

$$meq / L \left(\text{Conversion Factor} \right) = \frac{Valence}{Formula\,Wt.} \tag{7-10}$$

The conversion factors for each ion are given in the *Excel* spreadsheet for this problem. A more complete list can be found in the *Reference Book* Appendix.

The charge balance discussion from Problem 1 also applies to Problem 2. Thus, the percentage difference between cations and anions expressed in equation 7-4 should be less than 5% (Hem, 1985). Refer to this previous discussion regarding possible reasons for a charge imbalance.

The prevalent chemical character (PCC) is shorthand for expressing the predominant cation and anion pair in a water sample. It is determined by examining relative percentages of each cation or anion. The cation and anion pair with the highest percentage determines the prevalent chemical character. For example, one water could have a PCC of Ca-SO$_4$, whereas another water could have a PCC of Na-SO$_4$, and another of Na-Cl.

Carbonate Solubility

Of particular interest in this exercise is the dissolution and precipitation of calcite, dolomite, and gypsum in a regional carbonate aquifer. Each of these reactions will be explored in relation to the regional evolution in groundwater chemistry. The reactions of particular interest are as follows.

Calcite dissolution and precipitation:

$$CaCO_{3\ (s)} \Leftrightarrow Ca^{2+} + CO_3^{2-} \tag{7-11}$$

Dolomite dissolution and precipitation:

$$CaMg(CO_3)_{2(s)} \Leftrightarrow Ca^{2+} + Mg^{2+} + 2CO_3^{2-} \tag{7-12}$$

Gypsum dissolution and precipitation:

$$CaSO_4 \bullet 2H_2O_{(s)} \Leftrightarrow Ca^{2+} + SO_4^{2-} + 2H_2O \tag{7-13}$$

To help determine which direction these reactions will proceed in the environment, we employ the law of mass action (Freeze and Cherry, 1979; Hem, 1985; Fetter, 2001). This concept allows us to determine if a solution is oversaturated, at equilibrium, or undersaturated with respect to a given mineral phase. The general form is of an equilibrium reaction is expressed as follows.

$$cC + dD \Leftrightarrow xX + yY \qquad (7\text{-}14)$$

At equilibrium, the rate of the forward reaction (left to right) is equal to the rate of the backward reaction (right to left). The equilibrium constant (K), or in the case of solubility, the solubility product (K_{sp}), is expressed by the following equation. (It is important to differentiate between solubility and solubility product. Solubility refers to the amount of substance that dissolves under given conditions. Solubility product refers to the equilibrium constant expressed by the law of mass action. See Fetter (2001) or Hem (1985) for further discussion.)

$$K_{sp} = \frac{\text{Products}}{\text{Reactants}} = \frac{[X]^x [Y]^y}{[C]^c [D]^d} \qquad (7\text{-}15)$$

where $[X]$ is the thermodynamically effective concentration of the X ion and x is the number of moles of the chemical constituent. The K_{sp} is either experimentally determined or calculated from thermodynamic properties. Concentration values for pure liquid or solid are defined as 1. These equilibrium constants must be adjusted for the appropriate water temperature in the aquifer if it is initially reported relative to the standard temperature of 25°C. Another consideration is the other ions in a solution. They may limit the solubility of a given salt because of the common-ion effect. Therefore, analytical concentrations must be corrected for ionic strength and the effect of chemical complexation in a solution. For most natural waters, chemical activities must be determined before applying the law of mass action.

Be aware that these equilibrium calculations provide only the chemical boundary conditions of the system and not the actual chemical conditions. Simplifying assumptions—such as ignoring the kinetics of reactions, complex solid-liquid interfaces, additional reactions, and uncertainty in the calculated equilibrium constants—will all exert control on the actual chemical conditions.

Chemical Activities

The thermodynamically effective concentrations used in the law of mass action are also known as chemical activities. These chemical activities represent the concentrations of ions accounting for the nonidealities from electrostatic interaction among the ions in solution. The chemical activity for a given ionic species can be expressed with the following equation.

$$\alpha = \gamma m \qquad (7\text{-}16)$$

where

α is the chemical activity,
γ is the activity coefficient (M/mole), and
m is the molal concentration (mole/M).

Chemical activity coefficients are generally close to unity (~1.0) for dilute solutions and decrease as the salinity of the solution concentration increases. To determine the activity coefficient of an ion, we must first calculate the ionic strength of the solution. Ionic strength takes into account all of the ions in solution that may decrease the chemical activity of a particular ion. The ionic strength for a solution is expressed as follows.

$$I = \frac{1}{2} \sum m_i z_i^{\,2}$$

(7-17)

where

I is the ionic strength of the electrolyte solution (mole/M),
m_i is the molality of the ith ion (mole/M), and
z_i is the charge of the ith ion.

The activity coefficient of an individual ion can be determined using the extended Debye-Hückel equation for solutions with ionic strengths less than 0.1 M (TDS ≈ 5,000 mg/L).

$$\log \gamma_i = \frac{-A z_i^{\,2} \sqrt{I}}{1 + a_i B \sqrt{I}}$$

(7-18)

where

γ_i is the activity coefficient of the ionic species i,
z_i is the charge of the ionic species i,
I is the ionic strength,
A is a Debye-Hückel constant based on water temperature,
B is a Debye-Hückel constant based on water temperature, and
a_i is the effective diameter of the ionic species I (L).

Values for the Debye-Hückel constants A and B and the effective diameter parameter (a_i) can be found in Fetter (2001) or other general aqueous geochemistry textbooks. Values for these parameters, which are necessary to complete the problems, are given in the spreadsheet.

Ion Activity Product

Groundwater systems may be at a partial equilibrium state. Thus, some reactions such as dissolution-precipitation reactions may not be at equilibrium. The ion activity product (K_{iap}) is used to determine if a mineral is at saturation in a given water sample. Calculated values of K_{iap} for equilibrium reactions are compared with known solubility products (K_{sp}) for a given mineral. For the general dissolution reaction, cC ↔ xX + yY, the K_{iap} is calculated as follows with activity of the solid (α_C) equal to 1Y.

$$K_{iap} = \frac{(\alpha_X)^x (\alpha_Y)^y}{(\alpha_C)^c}$$

(7-19)

where

α_X is the chemical activity of the dissolution product X,

α_Y is the chemical activity of the dissolution product Y, and

α_C is the chemical activity of the dissolution reactant C (solid phase equals one).

Saturation Index

The saturation state of a mineral is commonly expressed in terms of the saturation index. This parameter allows for the comparison between the calculated ion activity product and the known solubility product.

$$SI = \log_{10} \frac{K_{iap}}{K_{sp}} \tag{7-20}$$

where

SI is the saturation index.

If SI = 0, then it is in equilibrium with the solution, whereas if SI < 0, then the mineral is undersaturated, and if SI > 0, then the mineral is oversaturated.

Because of the uncertainties inherit in the calculation of saturation indices, values in a small range near zero, 0 ± 0.5, are considered to be within the equilibrium range of a mineral (Deutsch, 1997).

Mineral Solubility

We will use reactions 7-11, 7-12, and 7-13 from this chapter to determine the ion activity product (equation 7-19) for each of the three minerals examined in this groundwater flow system. Each is expressed below.

$$K_{iap\,calcite} = \frac{\alpha_{Ca^{2+}} \; \alpha_{CO_3^{2-}}}{\alpha_{CaCO_3}} \tag{7-21}$$

$$K_{iap\,dolomite} = \frac{\alpha_{Ca^{2+}} \; \alpha_{Mg^{2+}} \left(\alpha_{CO_3^{2-}}\right)^2}{\alpha_{CaMg(CO_3)_2}} \tag{7-22}$$

$$K_{iap\,gypsum} = \frac{\alpha_{Ca^{2+}} \; \alpha_{SO_4^{2-}} \left(\alpha_{H_2O}\right)^2}{\alpha_{CaSO_4 \bullet 2H_2O}} \tag{7-23}$$

Commonly, the concentration of CO_3^{2-} is below the detection limit of the analytical method used to determine it, and non-detect (ND) is reported on the analytical concentration table for this anion. However, we need the activity of CO_3^{2-} to determine the K_{iap} for calcite. To do this, we use the activity of the HCO_3^- to calculate the activity of CO_3^{2-} as follows,

$$HCO_3^- \Leftrightarrow H^+ + CO_3^{2-} \qquad\qquad (7\text{-}24)$$

$$\frac{\alpha_{H^+}\,\alpha_{CO_3^{2-}}}{\alpha_{HCO_3^-}} = 10^{-10.49} \quad @\,10^\circ C \qquad\qquad (7\text{-}25)$$

We then solve for the activity of CO_3^{2-} by using the values of pH = –log [H^+] for the activity of H^+ and the calculated value of the activity of HCO_3^-.

Excel Tips

- SUM() sums the cells designated within the parentheses.
- Cell anchor: $column$row anchors the cell (column, row) in a formula. Performing a relative copy will maintain the anchored cell within the formula and change only the other non-anchored cells in the formula (e.g., =$D$12+D2 will anchor cell D12 and not cell D2 during a relative copy).
- AVERAGE(): returns the average of the cells designated within the parentheses.
- SQRT() is an intrinsic function in *Excel* that returns the square root of the quantity inside the parentheses.
- Exponent symbol: ^ raises a number to the designated power (e.g. 3^2 = 9).
- Logarithm relation: if Y = $\log_{10}$ X, then 10^Y = X.
- Remember to save your work often. *Excel* has an AutoSave function available under the Tools menu that may be an Add-In function with some versions.

Parameter	Definition
a_i	Effective diameter of the ionic species I (L)
A	Debye-Hückel constant based on water temperature
B	Debye-Hückel constant based on water temperature
α	Chemical activity (dimensionless)
I	Ionic strength (mole/M)
K_{iap}	Ion activity product
K_{sp}	Solubility product
m	Molal concentration (mole/M)
m_i	Molality of the ith ion (mole/M)
SI	Saturation index
γ	Activity coefficient (M/mole)
z_i	Charge of the ith ion

Table 7.4 Parameter Definition Table – Problem 2

Problem 3

GEOCHEMICAL EVALUATION OF INDUCED INFILTRATION FROM MIXING SURFACE WATER AND GROUNDWATER

South Well Field, Columbus, Ohio

Overview

Several of the concepts in this exercise build on those described previously in this chapter. These include the conversion of concentrations in milligrams per liter to milliequivalents per liter (equations 7-3 and 7-10), the calculation of total dissolved solids, prevalent chemical character, and the determination of charge balance. The reader is referred to the *Reference Book* and the references cited for additional background information and concepts.

Induced Infiltration

Water supply wells are commonly located next to rivers or streams to intercept a portion of the groundwater that normally would discharge into the river and to induce surface water to flow into the aquifer and then laterally into the well. A pumping well near a stream will create a depression in the potentiometric surface that captures water from the surrounding area and can reverse local flow directions (see Principles & Concepts in Chapter 4). Water that leaves the stream as a result of this change in the hydraulic gradient is referred to as induced infiltration (Winter et al., 1998). The amount of induced infiltration is a function of several parameters such as streambed permeability, stream stage, pumping rate, hydraulic gradient, and the hydraulic characteristics of the geologic materials between the well and the stream.

There are several ways to quantify the amount of induced infiltration that occurs near shallow production wells. Some of these methods were discussed in Chapter 4 under groundwater and surface water interactions. For example, stream gaging techniques were used to determine if the Aberjona River in eastern Massachusetts was gaining or losing water as a result of pumping at Woburn wells G and H. In addition to this technique, streambed permeability studies, mini-piezometers, tracer studies, flow models, and geochemical studies are used to quantify the amount of groundwater and surface water interaction (Fetter, 2001).

This exercise uses geochemical techniques to estimate the amount of mixing between two chemically distinct waters. In this application, the mixing is assumed to be linear between two end-members that are natural waters and are chemically conservative (i.e., no geochemical reactions are occurring to change the types of ions dissolved in the system). The example problem, shown in Figure 7.5, represents mixing between aquifer water and river water to produce the mixture of water found in a local well. Three samples from each locality (river, aquifer, and well) were collected and relative concentrations determined in the laboratory. The sodium concentrations in meq/L were plotted versus the chloride concentrations in meq/L. Assuming a simple, linear mixing relation between the end-member waters, we can calculate the percentage of each end-member water in the final mixture. The relative amount of each end-member water is calculated by first measuring the distance between the end-member waters on the plot. Then, measure the distance between one end-member water and the mixed water sample on the plot. In the example problem in Figure 7.5, the relative distances are recorded at the top of the plot. The figure also shows a graduated scale used to determine the amount of aquifer water present in the well water. Thus, on the basis of the relative distance on the sodium versus chloride plot, we estimate that approximately 38% of the water sampled in the wells originated as aquifer water, which means that the remaining 62% of the well water is derived from the river.

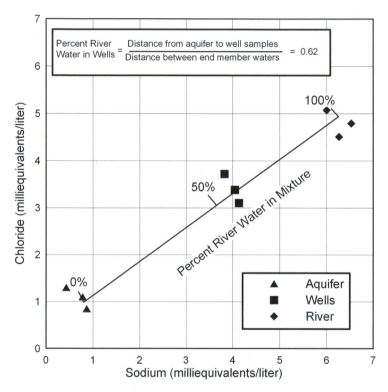

Figure 7.5. Graphical representation of the degree of mixing in sampled well water between two end-member waters; aquifer water is lower in sodium and chloride, and river water is higher in sodium and chloride.

It is commonly difficult to distinguish among mechanisms that may cause similar changes in water chemistry, particularly mixing, which is a widespread phenomenon. However, post-mixing reactions such as dissolution, precipitation, and ion exchange are common in natural waters. Assumptions that all ions remain in solution along a linear mixing line cannot be accepted unconditionally (Cheng, 1988).

Graphic Representation of Major Ions

Chemical analysis of water samples generally produces large quantities of data that can be displayed in tables or as graphs. These graphs are prepared in a variety of ways to aid in the interpretation of the chemical analysis results. Many of these graphical methods represent percentages of cation concentrations and percentages of anion concentrations, as in bar charts, pie diagrams, Stiff diagrams, and Piper diagrams (see Schwartz and Zhang, 2003).

Of particular use in completing the exercise are Piper diagrams, which are composed of two trilinear diagrams (plots with three axes) and one diamond-shaped mixing field. The trilinear diagram on the left-hand side represents the cation concentrations and the trilinear diagram on the right-hand side represents the anion concentrations. The diamond-shaped mixing field between the two trilinear diagrams allows the graphical interpretation of the mixing of two or more end-member waters (Figure 7.6). Water composition is expressed in percentages of the total cations and total anions. For the cation trilinear portion, Mg, Ca, and Na+K are each given one axis with 100% at each corner of the triangle. For the anion trilinear portion, Cl, SO_4, and CO_3+HCO_3 are also each given one axis with 100% at the corners. The diamond-shaped mixing field represents both cation and anion concentrations as projected from the cation and anion trilinear portions (Figure 7.6).

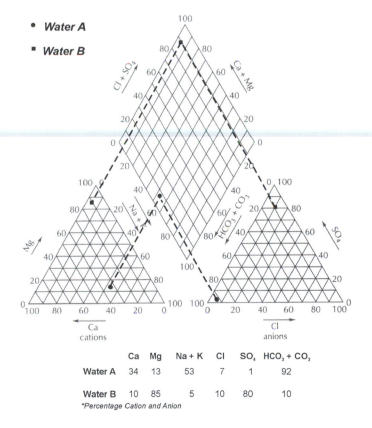

	Ca	Mg	Na + K	Cl	SO_4	$HCO_3 + CO_3$
Water A	34	13	53	7	1	92
Water B	10	85	5	10	80	10

*Percentage Cation and Anion

Figure 7.6. Piper diagram showing two water samples plotted in the cation and anion trilinear portions and projected into the mixing diamond. Relative percentage of cations and anions for each water sample is listed at the bottom of the figure.

Excel Tips

- AVERAGE(): returns the average of the cells designated within the parentheses.
- Cell anchor: $column$row anchors the cell (column, row) in a formula. Performing a relative copy will maintain the anchored cell within the formula and change only the other non-anchored cells in the formula (e.g., =$D$12+D2 will anchor cell D12 and not cell D2 during a relative copy).
- Other intrinsic *Excel* functions are listed and explained by clicking the f_x button.
- Remember to save your work often. *Excel* has an AutoSave function available under the Tools menu that may be an Add-In function with some versions.

Parameter	*Definition*
$C_{mg/L}$	Solute concentration in milligrams per liter (M/V)
C_{ppm}	Solute concentration parts per million (M/M)
C_M	Solute concentration in molarity (mole/V)
$C_{meq/L}$	Solute concentration in milliequivalents per liter (milliequivalents/V)
C_H	Concentration of $CaCO_3$ hardness in milligrams per liter
FW	Formula weight in grams per mole (M/mole)
TDS	Total dissolved solids in milligrams per liter (M/V)
V	Valence charge of an ion
ρ_w	Density of the water in grams per cubic centimeter (M/V)

Table 7.5 Parameter Definition Table – Problem 3

Appendix A
Unit Conversion Table

LENGTH

1 in (inch)	=	2.54 centimeters
1 ft (foot)	=	0.3048 meters
	=	5,280. feet
1 mi (mile)	=	1,609. meters
	=	1.609 kilometers

VOLUME

	=	231 cubic inches
	=	0.1337 cubic feet
1 U.S. gallon	=	3.785 liters
	=	0.003785 cubic meters
1 Mgal (million U.S. gallons)	=	3.069 acre-feet
	=	1,728 cubic inches
1 ft^3 (cubic foot)	=	7.481 U.S. gallons
	=	0.0283 cubic meters
	=	43,560 cubic feet
1 acre-ft (acre-foot)	=	1,234. cubic meters

VELOCITY AND GRADIENT

1 mi/hr (miles per hour)	=	1.4667 ft/s (feet per second)
	=	0.4470 m/s (meters per second)
1 ft/mi (feet per mile)	=	0.0001894 ft/ft (feet/foot)

FLOW RATE

1 ft^3/s (cubic feet per second)	=	448.8 gal/min (gallons per minute)
	=	0.6463 Mgal/d (million gallons per day)
	=	1.984 acre-ft/d (acre-feet per day)
	=	0.0283 m^3/s (cubic meters per second)

HYDRAULIC CONDUCTIVITY

1 ft/d (feet per day)	=	7.481 gal/d/ft^2 (U.S. gallons per day per square foot)
	=	0.305 m/d (meters per day)
	=	0.000353 cm/s (centimeters per second)

TRANSMISSIVITY

1 ft^2/d (square feet per day)	=	7.481 gal/d/ft (U.S. gallons per day per foot)
	=	0.0929 m^2/d (square meters per day)

ANGLES

1 radian	=	57.30 degrees
1 degree	=	0.01746 radians

CONCENTRATIONS

1 mg/L (milligram per liter)	=	1,000 µg/L (micrograms per liter)
1 ppm (part per million)	=	1,000 ppb (parts per billion)

Appendix B
Rock and Sediment Hydraulic Conductivity Values

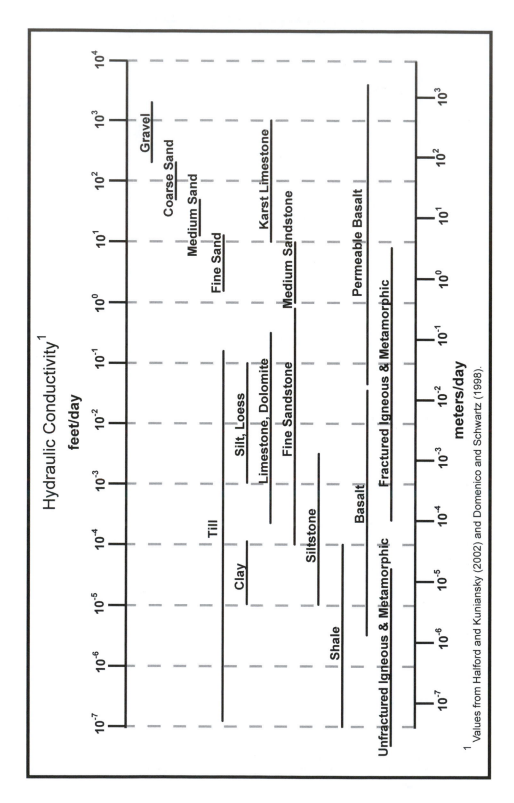

Appendix C
Rock and Sediment Porosity
and Specific Yield Values

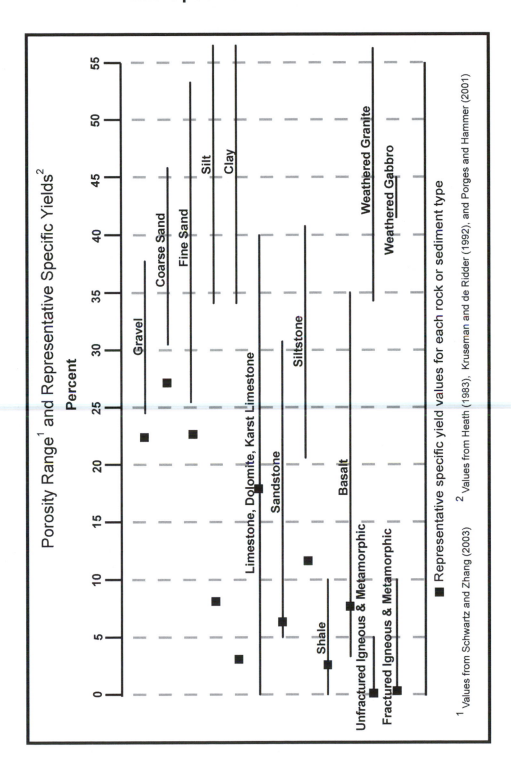

Appendix D
Milliequivalent Conversion Factors[1]

Element and Reported Species	Gram Formula Weight	Conversion Factor from mg/L to meq/L
Calcium (Ca^{2+})	40.08	0.04990
Iron (Fe^{2+})	55.85	0.03581
Iron (Fe^{3+})	55.85	0.05372
Magnesium (Mg^{2+})	24.31	0.08229
Manganese (Mn^{2+})	54.94	0.03640
Manganese (Mn^{4+})	54.94	0.07281
Potassium (K^+)	39.10	0.02558
Sodium (Na^+)	22.99	0.04350
Bicarbonate (HCO_3^-)	61.02	0.01639
Carbonate (CO_3^{2-})	60.01	0.03333
Chloride (Cl^-)	35.45	0.02821
Fluoride (F^-)	19.00	0.05264
Nitrate (NO_3^-)	62.01	0.01613
Phosphate (PO_4^{3-})	94.97	0.03159
Phosphate (HPO_4^{2-})	94.97	0.02084
Phosphate ($H_2PO_4^-$)	94.97	0.01031
Sulfate (SO_4^{2-})	96.06	0.02082

[1]Data from Hem (1985)

Concentration in milligrams/liter X conversion factor = concentration in milliequivalents per liter

REFERENCES

Abriola, L.M.and G.F. Pinder. 1982. Calculation of velocity in three space dimensions from hydraulic head measurements. *Ground Water*, vol. 20, no. 2, p. 205-213.

ASTM-D422. 2002. *Standard Test Method for Particle-Size Analysis of Soils.* American Society for Testing and Materials (ASTM).

Back and Hanshaw. 1965. *Chemical geohydrology.* Adv. Hydrosci., vol. 1, p. 49-109.

Bair, E.S., 2001. Models in the Courtroom, M.G. Anderson and P.D. Bates eds., in *Model Validation: Perspectives in Hydrological Science,* John Wiley & Sons, Ltd., London, p. 57-77.

Bair, E.S. and M.A. Metheny. 2002. Remediation of the Wells G & H Superfund Site, Woburn, Massachusetts, *Ground Water*, vol. 40, no. 6, p. 657-668.

Bair, E.S., R.A. Sheets, and S.M. Eberts. 1990. Particle-tracking analysis of flow paths and traveltimes from hypothetical spill sites within the capture zone of a well field. *Ground Water*, vol. 28, no. 6, p. 884-892.

Bradbury, K.R. and M.A. Muldoon, 1990. Hydraulic conductivity determinations in unlithified glacial and fluvial materials. in *Ground Water and Vadose Zone Monitoring,* ASTM Special Technical Publication 1053, D.M. Nielson and A.I. Johnson, editors, p. 138-151.

Brasier, F.M. and B.J. Kobelski. 1996. Injection of industrial wastes in the United States. in *Deep Injection Disposal of Hazardous and Industrial Waste.* J.A. Apps and C. Tsang editors. p. 1-8.

Butler, J.J. 1998. *The Design, Performance, and Analysis of Slug Tests.* Boca Ranton, Florida: Lewis Publishers, 252 p.

Butler, J.J. and C.D. McElwee. 1990. Hydrogeologic characterization of hazardous waste sites. Contributions to Kansas Water Resources Research Institute, vol. 283, 11 p.

Bouwer, H. 1989. The Bouwer and Rice slug test--an update. *Ground Water*, vol. 27, no. 3, p. 304-309.

Bouwer, H. and R.C. Rice. 1976. A slug test method for determining hydraulic conductivity of unconfined aquifers with completely or partially penetrating wells, *Water Resources Research*, vol. 12, no. 3, p. 423-428.

Chamberlin, T.C. 1885. Requisition and qualifying conditions of artesian wells. U.S. Geological Survey Annual Report 5: Frontispiece.

Chebotarev, I. I. 1955. Metamorphism of natural waters in the crust of weathering, *Geochimica et Cosmochimica Acta*, vol. 8, no. 1-4, p. 22-48.

Cheng, S. 1988. Trilinear diagram revisited: application, limitation, and an electronic spreadsheet program. *Ground Water*, vol. 26, no. 4, p. 505-510.

Childress, J.O., R.A. Sheets, and E.S. Bair. 1991. Hydrology and water quality near the South Well Field, southern Franklin County, Ohio, with emphasis on the simulation of ground-water flow and transport of Scioto River. U.S. Geological Survey. *Water-Resources Investigations* 91-4080, 78 p.

Chirlin, G.R. 1990. The slug test: The first four decades. *Ground Water Management*, vol. 1, p. 365-381.

Chute, N.E. 1959. Glacial geology of the Mystic Lakes-Fresh Pond area, Massachusetts. U.S. Geological Survey *Bulletin* 8755-531X ; B 1061-F, p. 187-216.

Commonwealth of Massachusetts. 1922. Pollution Survey of the Aberjona River, 1921-1922, Division of Fisheries and Wildlife, Westboro, Massachusetts, 160 annotated photographs.

Cooper, H.H. Jr., J.D. Bredehoeft, and I.S. Papadopulos. 1967. Response of a finite-diameter well to an instantaneous charge of water. *Water Resources Research*, Vol. 3, No. 1, p. 263-269.

Cooper, H.H. Jr. and J.D. Jacob. 1946. A generalized graphical method for evaluating formation constants and summarizing well-field history. *Transactions*, American Geophysical Union, vol. 27, p. 526-534.

Cunningham, W.L. 1992. Hydrogeology and simulation of transient ground-water flow at the South Well Field, Columbus, Ohio. Master's Thesis, Department of Geological Sciences, The Ohio State University, 154 p.

Cunningham, W.L, E. S. Bair, W. P. Yost. 1996. Hydrogeology and simulation of ground-water flow at the South Well Field, Columbus, Ohio. *Water-Resources Investigations* - U. S. Geological Survey 0092-332X ; WRI 95-4279, 56 p.

Davis, S.N. and R.J.M. De Wiest. 1966. *Hydrogeology*. John Wiley & Sons, New York, 463 p.

Delong, R.M. and G.W. White. 1963. Geology of Stark County. Ohio Division of Geological Survey. Bulletin 61, 209 p.

Deutsch, W.J. 1997. *Groundwater Geochemistry -- Fundamentals and Applications to Contamination.* New York: Lewis Publishers, 221 p.

Domenico, P.A. and F.W. Schwartz. 1998. *Physical and Chemical Hydrogeology.* New York: John Wiley and Sons, Inc., 506 p.

Dunne, Thomas and Luna Leopold. 1978. *Water in Environmental Planning.* Chapter 10, Calculation of Flood Hazard. San Francisco: W.H. Freeman and Company, 818 p.

Eberts S.M. and L.L. George. 2000. Regional Ground-Water Flow and Geochemistry in the Midwestern Basins and Arches Aquifer system in Parts of Indiana, Ohio, Michigan, and Illinois. U.S. Geological Survey *Professional Paper* 1423-C.

Environmental Protection Agency. 2002. Protecting drinking water through underground injection control. *Drinking Water Pocket Guide #2.* EPA 816-K-02-001.

Environmental Protection Agency. 2004. Hazardous Waste Goes Deep. Region 6 District Underground Injection Control Program <http://www.epa.gov/region6/6xa/uic_deep.htm>

Fetter, C.W. 2001. *Applied Hydrogeology.* New Jersey: Prentice Hall, 598 p.

Freeze, R.A. and J.A. Cherry. 1979. *Groundwater.* Prentice-Hall, Englewood Cliffs, N.J., 604 p.

Goldthwait, R. P., J.L. Forsyth, G. W. White. 1961. Glacial Map of Ohio, Ohio Division of Geological Survey.

Grubb, S. 1993. Analytical model for estimation of steady-state capture zones of pumping wells in confined and unconfined aquifers. *Ground Water*, vol. 31, no. 1, p. 27-32.

Harr, J. 1995. *A Civil Action.* New York: Random House, 500 p.

Harsh, J.F. and R.J. Laczniak. 1990. Conceptualization and analysis of ground-water flow system in the Coastal Plain of Virginia and adjacent parts of Maryland and North Carolina. U.S. Geological Survey *Professional Paper* 1404-F, 100 p.

Hazen, A. 1911. Discussion: Dams on sand foundations. *Transaction*s, American Society of Civil Engineers, vol. 73, p. 199.

Hem, J. D. 1985. Study and interpretation of the chemical characteristics of natural water. U. S. Geological Survey, *Water-Supply Paper* 2254, 264 p.

Hess, K.M. 1988. National Water Summary 1986, Sewage plume in a sand and gravel aquifer: a case study: U.S. Geological Survey *Water-Supply Paper* 2325, p. 87-92.

Hubbert, M.K. 1940. The theory of ground-water motion. *Journal of Geology*, vol. 48, no. 8, p. 785-944.

Hvorslev, M.J., 1951. Time Lag and Soil Permeability in Ground-Water Observations, Bull. No. 36, Waterways Exper. Sta. Corps of Engrs, U.S. Army, Vicksburg, Mississippi, p. 1-50.

Hyder, Z., J.J. Butler, Jr., C.D. McElwee, and W. Liu, 1994. Slug tests in partially penetrating wells, *Water Resources Research*, vol. 30, no. 11, p. 2945-2957.

Jacob, C.E. 1944. Notes on determining permeability by pumping tests under watertable conditions. U.S. Geological Survey, *Open File Report.*

Joseph, J.A. 1995. A Hydro-Geochemical Assessment of Groundwater Flow and Arsenic Contamination in the Aberjona River Sub-basin, M.S. thesis, Department of Civil and Environmental Engineering, MIT.

Kruseman, G.P. and N.A. de Ridder. 1990. Analysis and evaluation of pumping test data. Netherlands: International Institute for Land Reclamation and Improvement, 377 p.

Lahm, T.D. 1997. Numerical analysis of capture-zone geometries for a partially penetrating well and hydrodynamics of salinity-derived variable-density flow in regionally extensive aquifers in Midcontinent sedimentary basins. Ph.D. Dissertation. Department of Geological Sciences. The Ohio State University, 144 p.

LeBlanc, Dennis R. 1984. Sewage Plume in a Sand and Gravel Aquifer, Cape Cod, Massachusetts. Washington: U.S. Geological Survey. *Water-Supply Paper* 2218, 28 p.

Lewelling, B.R. 1988. Potentiometric surface of the intermediate aquifer system, west-central Florida. May 1987: U.S. Geological Survey *Open-File Report* 87-705, scale 1:1,000,000, 1 sheet.

Leopold, L.B. 1968. Hydrology for urban land planning – A guidebook on the hydrologic effects of urban land use. Washington: U.S. Geological Survey. Geologic Survey *Circular* 554, 18p.

Littin, Gregory R. 1999. Monitoring the effects of ground-water withdrawals from the N Aquifer in the Black Mesa area, Northeastern Arizona, U.S. Geological Survey Fact Sheet FS-064-99.

Lloyd, O.B. and W.L. Lyke. 1995. *Ground Water Atlas Of The United States; Illinois, Indiana, Kentucky, Ohio, Tennessee,* HA 730-K U.S. Geological Survey.

Meng, A.A. and J.F. Harsh. 1988. Hydrogeologic framework of the Virginia Coastal Plain.

U.S. Geological Survey *Professional Paper* 1404-C, 128 p.

Metheny, M.A. 1998. Hydrogeologic framework and numerical simulation of flow to well G & H at Woburn, Massachusetts, unpublished M.S. thesis, Department of Geological Science, The Ohio State University, 247 p.

Metheny, M. and E.S. Bair. 2001. The Science Behind "A Civil Action"; Hydrogeology of the Aberjona River, Wetland, and Woburn Wells G and H. in *Guidebook for Geological Field Trips in New England.* Geological Society of America, p. D1-D25.

Metheny, M. 2004. Evaluation of groundwater flow and contaminant transport at the Wells G&H Superfund Site, Woburn, Massachusetts, from 1960 to 1986 and estimation of TCE and PCE concentrations delivered to Woburn residences. Doctor of Philosophy, Ohio State University, Geological Sciences, 367 p.

Miller, James A. 1990. *Ground Water Atlas Of The United States; Alabama, Florida, Georgia, South Carolina,* HA 730-G U.S. Geological Survey.

Moench, A.F. and P.A. Hsieh. 1985. Analysis of slug test data in a well with finite-thickness skin. Mem. IAH Int. Congr. Hydrogeol. Rocks Low Permeability, vol. 17, no. 2, p. 17-29.

Myette, C.F., D.G. Johnson, J.C. Olimpio. 1987. Area of influence and zone of contribution to superfund site wells G and H, Woburn, Massachusetts. U.S. Geological Survey, *Water-Resources Investigations 87-4100, 86 p.*

Norris, S.E. and R.E. Fidler. 1973. Availability of water from limestone and dolomite aquifers in southwest Ohio and the relation of water quality to the regional flow system. U.S. Geological Survey. *Water-Resources Investigations* 17-73, 42 p.

Norris, S.E. and A.M. Spieker. 1966. Ground-water resources of the Dayton area, Ohio. U.S. Geological Survey *Water-Supply Paper* 0886-9308 ; W 1808, 167 p.

Ogata, A. and R.B. Banks. 1961. A solution of the differential equation of longitudinal dispersion in porous media. U.S. Geological Survey *Professional Paper* 411-A, p. A1-A7.

Olcott, P.G. 1995. *Ground Water Atlas Of The United States; Connecticut, Maine, Massachusetts, New Hampshire, New York, Rhode Island, Vermont,* HA 730-M, U.S. Geological Survey.

Peabody Western Coal Company. 1999. A Three-Dimensional Flow Model of the D and N Aquifers, Black Mesa Basin, Arizona. HSI GeoTrans and Waterstone Report Volume I.

Peters, J.G. 1987. Description and comparison of selected models for hydrologic analysis of ground-water flow, St. Joseph River basin, Indiana. U.S. Geological Survey *Water-Resources Investigations* - WRI 86-4199, 125 p.

Plummer, L.N. 1977. Defining reactions and mass transfer in part of the Floridan Aquifer. *Water Resources Research*, vol. 13, no. 5, p. 801-812.

Payne, C. 1992. Delineation of traveltime-related capture zones of the municipal wells at Circleville, Ohio. Unpublished Thesis (M.S.)--Ohio State University, 240 p.

Rantz, S.E. 1982. Measurement and computation of streamflow; Volume 1, Measurement of stage and discharge; Volume 2, Computation of discharge. *U.S. Geological Survey, Water-Supply Paper* 0083-1131, 631 p.

Renken, R.A. 1998. *Ground Water Atlas of The United State; Arkansas, Louisiana, Mississippi*

U.S. Geological Survey HA 730-F.

Ritter, Dale, Craig Kochel, and Jerry Miller. 1995. *Process Geomorphology*. Wm. C. Brown Publishers, 543 p.

Robson, S.G. and E.R. Banta. 1995. *Ground Water Atlas of the United States, Arizona, Colorado, New Mexico, Utah*. U.S. Geological Survey HA 730-C.

Rutledge, A.T. 1998. Computer programs for describing the recession of ground-water discharge and for estimating mean ground-water recharge and discharge from streamflow data – update: *U.S. Geological Survey Water-Resources Investigations* 98-4148. 43 p.

http://water.usgs.gov/pubs/wri/wri984148/

Rye, R.O., W. Back, B.B. Hanshw, C.T. Rightmire, and F.J. Pearson. 1981. The origin and isotopic composition of dissolved sulfide in groundwater from carbonate aquifers in Florida and Texas. *Geochimica et Cosmochimica Acta*, vol. 45, p. 1941-1950.

Sanders, L.L. 1998. *A Manual of Field Hydrogeology*. Prentice Hall, New Jersey, 381 p.

Schwartz, F.W. and H. Zhang. 2003. *Fundamentals of Ground Water*. John Wiley and Sons, Inc. New York, 583 p.

Singh, Vijay. 1987. *Regional Flood Frequency Analysis*. Proceedings of the International Symposium of Flood Frequency and Risk Analysis, 14-17 May 1986, Louisiana State University, Baton Rouge, USA. Boston: Kluwer Academic Publishers, 400 p.

Smajstrla A.G. and D.Z. Haman. 2002. Irrigated Acreage in Florida: A Summary through 1998. CIR 1220. Cooperative Extension Service of University of Florida, Institute of Food and Agricultural Sciences (UF/IFAS).

Sminchak, J.R. 1997. Seasonal ground-water levels in Ohio. Unpublished Master's Thesis, The Ohio State University, 101 p.

Storck, R. J. and J.P. Szabo. 1991. Lithofacies and mineralogy of the late Wisconsinan Navarre Till in Stark and Wayne counties, Ohio. *Ohio Journal of Science*, vol. 91, no. 1, p. 90-97.

Stout, W., K. Ver Steeg, and G.F. Lamb. 1943. Geology of Water in Ohio. Ohio Geological Survey Bulletin 44, 694 p.

Strobel, M. L. and E.F. Bugliosi. 1991. Areal extent, hydrogeologic characteristics, and possible origins of the carbonate rock Newburg zone (Middle-Upper Silurian) in Ohio: *Ohio Journal of Science*, vol. 91, no. 5, p. 209-215.

Tarr, J.A. 1987. History of pollution in Woburn, Massachusetts, report prepared for W.R. Grace & Co., included as Appendix B to GeoTrans, 1987.

Theis, C.V. 1935. The lowering of the piezometer surface and the rate and discharge of a well using ground-water storage. *Transactions,* American Geophysical Union, vol. 16, p. 519-524.

Todd, D.K. 1980. *Groundwater Hydrology*. 2nd ed. John Wiley & Sons, New York, 535 p.

Touchstone Pictures. 1998. *A Civil Action*. Buena Vista Studios.

Toth, J.A. 1962. A theory of ground-water motion in small drainage basins in central Alberta, Canada. *Journal of Geophysical Research*, vol. 67, no. 11, p. 4375-4387.

Toth, J.A. 1963. A theoretical analysis of ground-water flow in small drainage basins. *Journal of Geophysical Research*, vol. 68, no. 16, p. 4795-4811.

Trapp, H. and M.A. Horn. 1997. *Ground Water Atlas Of The United State; Delaware, Maryland, New Jersey, North Carolina, Pennsylvania, Virginia, West Virginia*
U.S. Geological Survey HA 730-L.

Truini, M. and E.T. Blakemore. 2003. Ground-Water, Surface-Water, and Water-Chemistry Data, Black Mesa Area, Northeastern Arizona – 2003-03. U.S. Geological Survey, *Open-File Report* 03-503, 54 p.

Truini, M. and S.A.Longsworth. 2003. Hydrogeology of the D Aquifer and Movement and Ages of Ground Water Determined from Geochemical and Isotropic Analyses, Black Mesa Area, Northeastern Arizona. U.S. Geological Survey, *Water-Resources Investigations* 03-4189, 48 p.

U.S. Army Corp of Engineers, 2004, Environmental Glossary
<http://www.lrb.usace.army.mil/fusrap/glossary-gh.htm>

U.S. Bureau of Census. 2000. U.S. National Census.

U.S. Water Resources Council. 1981. Guideline For Determine Flood Frequency. Bulletin 17B of the Hydrology Subcommittee, Interagency Agency Advisory Committee. USGS. Washington, D.C., 194 p.

Weight, W.D. and J.L. Sonderegger. 2001. *Manual of Applied Field Hydrogeology*. McGraw-Hill. 608 p.

White, G.W. 1982. Glacial Geology of Northeastern Ohio. Ohio Geological Survey Bulletin 68, 75 p.

Winter, T.C., J.W. Harvey, O.L. Franke, and W.M. Alley. 1998. Ground Water and Surface Water, A Single Resource. U.S. Geological Survey *Circular* 1139, 79 p.

Wolansky, R.M. and M.A.Corral, Jr. 1985. Aquifer tests in west-central Florida, 1952-76: U.S. Geological Survey *Water-Resources Investigations* 84-4044, 127 p.

E. Scott Bair and Terry D. Lahm
Practical Problems in Groundwater Hydrology CD 1e
0-13-147589-4
© 2006 Pearson Education, Inc.
Pearson Prentice Hall
Pearson Education, Inc.
Upper Saddle River, NJ 07458
Pearson Prentice Hall™ is a trademark of Pearson Education, Inc.

System requirements

Windows - Minimum configuration
*Processor; Intel Pentium II processor or equivalent
*RAM; 64 megs of RAM for Windows 98, 128 megs of RAM for Windows 2000, 128 megs of RAM for Windows XP
*Operating System; Windows 98, 2000, XP - please note that this CD is not designed to function on Windows ME and NT.
*Monitor Resolution: 1024x768
*Mouse or other pointing device
*Required Hardware; CD-ROM drive, Graphic card capable of 1024x768 resolution and 16 bit color
*Internet Connection Not Required
Macintosh – Minimum configuration
*Processor; G3 processor
*RAM; 128 megs of RAM
*Operating System: OS 10.1.5, 10.2.6, 10.3 or higher
*Monitor Resolution: 1024x768
*Mouse or other pointing device
*Required Hardware; CD-ROM drive, Graphic card capable of 1024x768 resolution and 16 bit color
*Internet Connection Not Required
For Both Platforms Software Required;
Microsoft Excel® is required to view, complete, and print the worksheets. All Excel content was created using Excel 2000. Users must have Excel 2000 or a compatible version to properly view the Excel Worksheets. Pearson does not support Microsoft Excel®. Please contact Microsoft for assistance. This CD is designed for stand-alone use only. It is not designed or licensed to run on a network.

Getting started
Windows
Inserting this CD-ROM will automatically start the application, if you need to manually start the application follow the steps below:

When the CD is inserted it should run at startup. If not, navigate to the contents of the CD and double click on the GWHydrology.exe icon to launch the program.
Macintosh
Inserting this CD-ROM will NOT automatically start the application, to manually start the application follow the steps below: When the CD is inserted it will show up on your desktop. Open the contents of the CD and double click on the Groundwater Hydrology icon to launch the program.

CD-ROM content
Practical Problems in Groundwater Hydrology application
Excel worksheets
Read Me

Known issues
Certain Worksheets within this product require that you run Macros within the Excel program. If you are presented with a message regarding Excel Macros, please allow the Macros to run. Certain Excel Worksheets will not function correctly if you reject the running of Macros.
If your computer system has multiple CD or DVD drives, you will need to access this product from the primary CD drive. If you experience problems with a certain CD or DVD drive on your system, please try another drive.
If you have just installed the Excel program on your computer, you must open an Excel spreadsheet before accessing Excel through the CD. Failing to do so will result in a failure of the Excel Worksheets present on this CD.
For users of Excel 2000, certain Excel Worksheets within this product require that you install an Excel Add-In titled - Analysis ToolPack. Please see your Excel documentation or contact Excel support for details on how to perform this installation.

Support information
If you are having problems with this software, call (800) 677-6337 between 8:00 a.m. and 8:00 p.m. EST, Monday through Friday, and 5:00 p.m. through 12:00 a.m. EST on Sundays. You can also get support by filling out the web form located at: http://247.prenhall.com/mediaform
Our technical staff will need to know certain things about your system in order to help us solve your problems more quickly and efficiently. If possible, please be at your computer when you call for support. You should have the following information ready:
- Textbook ISBN
- CD-Rom/Diskette ISBN
- Corresponding product and title
- Computer make and model
- Operating System (Windows or Macintosh) and Version
- RAM available
- Hard disk space available
- Sound card? Yes or No
- Printer make and model
- Network connection
- Detailed description of the problem, including the exact wording of any error messages.
NOTE: Pearson does not support and/or assist with the following:
- third-party software (i.e. Microsoft including Microsoft Office Suite - including Excel, Apple, Borland, etc.)
- homework assistance
- Textbooks and CD-ROMs purchased used are not supported and are non-replaceable. To purchase a new CD-Rom contact Pearson Individual Order Copies at 1-800-282-0693.